PROTEIN BIOCHEMISTRY, SYNTHESIS, STRUCTURE AND CELLULAR FUNCTIONS

ADVANCES IN PROTEIN FOLDING RESEARCH

PROTEIN BIOCHEMISTRY, SYNTHESIS, STRUCTURE AND CELLULAR FUNCTIONS

Additional books in this series can be found on Nova's website under the Series tab.

Additional e-books in this series can be found on Nova's website under the e-books tab.

PROTEIN BIOCHEMISTRY, SYNTHESIS, STRUCTURE AND CELLULAR FUNCTIONS

ADVANCES IN PROTEIN FOLDING RESEARCH

MATHEW HALE
EDITOR

nova publishers
New York

Library of Congress Cataloging-in-Publication Data

ISBN: 978-1-63483-593-0

Library of Congress Control Number: 2015947569

Published by Nova Science Publishers, Inc. † New York

CONTENTS

PREFACE

An important but relatively unknown protein modification is PHBylation. In this book, the authors review the physical properties and molecular characteristics of PHB, and discuss the influences of PHBylation on protein folding. In the second chapter, protein-folding mechanisms, namely the framework model and nucleation-condensation mechanism are discussed. The final chapter examines the fundamental physical law governing protein folding - the Thermodynamic Principle.

An important but relatively unknown protein modification is PHBylation, i.e., covalent attachment of short polymers (< 20 residues) of (R)-3-hydroxybutyrate (cPHB). PHB are water-insoluble and lipid soluble polymers in which hydrophobic methyl groups alternate with hydrophilic ester groups along a flexible linear chain. A cPHB molecule forms a single covalent bond at its ester end to a hydroxy amino acid of the protein (usually serine), and the remainder of the polymer may then form multiple noncovalent hydrophobic, hydrogen and coordinate bonds to neighboring amino acid residues. PHBylated proteins are present in all biological cells – prokaryotic and eukaryotic. They are primarily found in membranes and organelles, wherein they are frequently engaged in the binding and transport of ions or polyanions. Since the flexibility of the PHB backbone decreases markedly below physiological temperatures, the folding and the function(s) of cPHB-modified proteins are highly temperature-sensitive. In Chapter 1, the author review the physical properties and molecular characteristics of PHB, and discuss the influences of PHBylation on protein folding.

Analysis of Φ values provides information on structures of transition states along a folding pathway of a protein via measurement of effects of a mutation on folding kinetics. In Chapter 2, the author discuss the Φ values calculated for 27 proteins (54–128 amino acid residues long) by means of a simple statistical-mechanical model of protein folding (developed by us); the author compare these data with experimentally observed Φ values. In the model, three-dimensional structure of a protein was taken into consideration via inter-residue contact information of the native structure obtained from the Protein Data Bank. A free-energy profile of the wild-type protein was calculated by assigning a single value to interaction energies of the inter-residue contacts (regardless of the amino acid types) as an approximation. The same procedure was applied to a protein with a single point mutation but with slightly greater or smaller values assigned to the interaction energies between the mutated and other (unmutated) residues. The folding and unfolding rates were individually estimated from the free-energy profiles of the wild-type protein and mutant protein. The Φ value was then calculated for every residue from the differences in the folding or unfolding rates between the wild-type and the mutant. Despite the simplicity of the model, corresponding experimental Φ values were fairly well predicted for most proteins. The successful cases point to the importance of topology of the three-dimensional structure and support validity of assumptions of the model. Conversely, the cases of failure are suggestive of the need for interaction details such as amino acid types and interaction modes during the calculation and revealed limitations of the model. Protein-folding mechanisms, namely the framework model and nucleation-condensation mechanism, are also discussed in terms of the advantages and disadvantages of the proposed model.

The fundamental physical law governing protein folding is the Thermodynamic Principle: the peptide chain of a protein together with the protein's physiological environment determines the protein's native structure; and the native structure has the minimum Gibbs free energy among all conformations of the protein.

Beginning with a thought experiment indicating that protein folding should be considered in the single molecule level and there must be a general folding force that pushes a protein molecule fold along a folding path, Chapter 3 recalls the works leading to the thermodynamics principle. Arguments to support a single molecule understanding of the thermodynamic principle is given. After preparations including the topology of the conformational space $\mathcal{X}_{\mathfrak{A}}$ of a molecule $\mathfrak{A}$ and necessary thermodynamic and quantum statistics tools, a Gibbs free energy function $G(\mathbf{X};\mathfrak{E}_N;\mathfrak{A})$ is derived for a globular protein $\mathfrak{A}$, where $\mathbf{X} \in \mathcal{X}_{\mathfrak{A}}$, $\mathfrak{E}_N$ is the physiological environment of $\mathfrak{A}$. The derivation relies on making tailor made, nano sized tiny thermodynamic systems $\mathfrak{S}_{X,\mathfrak{E}}$, that contains only one protein molecule in the conformation **X** and a layer of material particles appear in the environment $\mathfrak{E}$.

Many natural phenomena in protein folding and protein structures can be explained with this Gibbs free energy function and its dependence on environment. Among them are a holistic view of protein structure which explains phenomenon such as the chameleon sequences; why just reducing hydrophobic surface area can produce secondary structures and hydrogen bonds; structure stability; allosteric proteins, docking, enzyme-substrate combination; and denaturation and refolding. The new unified explanations are at least as good as current, *ad hoc* ones, in some cases are much better. Testable predictions are given, first among them is the prediction of protein native structure from the only knowledge of the protein's amino acid sequence. It just becomes a pure mathematical problem of minimisation. Another testable prediction is that changed local environment causes amyloid and prion related diseases.

In: Advances in Protein Folding Research
Editor: Mathew Hale
ISBN: 978-1-63483-593-0

Chapter 1

Influence of Short-Chain Conjugated Poly-(R)-3-Hydroxybutyrates (cPHB) on Protein Folding

Rosetta N. Reusch

Department of Microbiology and Molecular Genetics, Michigan State University, East Lansing, Michigan, US

Abstract

An important but relatively unknown protein modification is PHBylation, i.e., covalent attachment of short polymers (<20 residues) of (R)-3-hydroxybutyrate (cPHB). PHB are water-insoluble and lipid soluble polymers in which hydrophobic methyl groups alternate with hydrophilic ester groups along a flexible linear chain. A cPHB molecule forms a single covalent bond at its ester end to a hydroxy amino acid of the protein (usually serine), and the remainder of the polymer may then form multiple noncovalent hydrophobic, hydrogen and coordinate bonds to neighboring amino acid residues. PHBylated proteins are present in all biological cells – prokaryotic and eukaryotic. They are primarily found in membranes and organelles, wherein they are frequently engaged in the binding and transport of ions or polyanions. Since the flexibility of the PHB backbone decreases markedly below physiological temperatures, the folding and the function(s) of cPHB-modified proteins are highly temperature-sensitive. Here we review the physical properties and molecular characteristics of PHB, and discuss the influences of PHBylation on protein folding.

INTRODUCTION

Polymers of the ketone body R-3-hydroxybutyrate (R-3HB), known as poly-(R)-3-hydroxybutyrates(PHB), are constituents of all biological cells, from the earliest protocells to the most advanced eukaryotic cells. The synthesis of PHB requires only acetate and reducing potential, both of which were available in the primordial environment. This suggests that they were among the earliest of the organic biopolymers, and preceded the formation of polynucleotides, polypeptides and polysaccharides. Long- chain PHB (>10,000 residues) are accumulated in cytoplasmic granules by only certain prokaryotes [1,2], but all biological cells contain medium-length PHB (<200 residues) in their membranes and organelles (mPHB), and possess proteins, primarily in membranes and organelles, that are modified by covalent attachment of short-chain PHB (<20 residues) (cPHB) [3-5]. The process of conjugating short-chain PHB to proteins is known as PHBylation. Here we examine the role(s) that PHBylation plays in the folding of cPHB-modified proteins.

STRUCTURE AND SYNTHESIS OF PHB

PHB are linear polymers of R-3-hydroxybutyrate – hydrophobic methyl groups alternate with hydrophilic ester groups along a flexible chain.

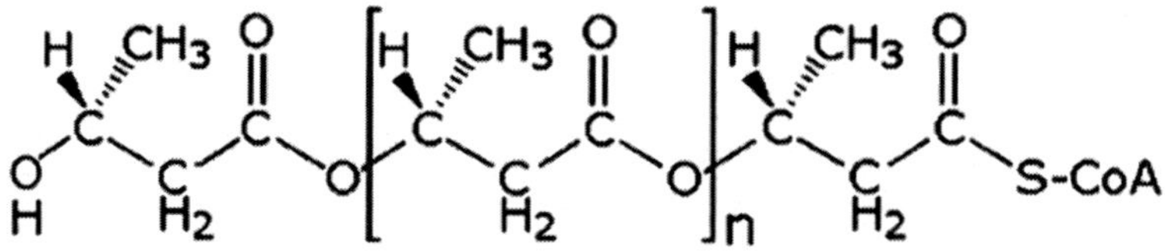

Figure 1. Structure of PHB. From Xian et al., 2007.

In both prokaryotes and eukaryotes, the synthesis of R-3-hydroxybutyrate (R-3-HB) begins with the condensation of two molecules of acetyl-CoA by a ketothiolase enzyme to form acetoacetyl-CoA. In prokaryotes (Figure 2 right), acetoacetyl-CoA is subsequently reduced by acetoacetyl-CoA reductase with NADPH to form R-3-hydroxybutyryl-CoA (R-3-HB-CoA), which is then polymerized by the enzyme PHB synthase to form long-chain PHB [6]. Enzymes that polymerize R-3-HB-CoA to mPHB have not yet been identified; however, Dai et al., [7] identified an enzyme, *ydcS*, that polymerizes R-3-HB-CoA to cPHB in the periplasm of *Escherichia coli*. In eukaryotes (Figure 2 left), acetoacetyl-CoA is condensed with a third acetyl-CoA by 3-hydroxy-3-methylglutaryl-CoA synthase (HMG synthase) to form 3-hydroxy-3-methylglutaryl-CoA (HMG-CoA) (Figure 2 right). The enzyme HMG-lyase then catalyzes the decomposition of HMG-CoA to form acetoacetate and acetyl-CoA, and acetoacetate is further reduced with NADH by R-3-HB dehydrogenase to form R-3-HB [8]. Enzymes that polymerize R-3-HB or its CoA ester to mPHB or cPHB have not yet been identified in eukaryotes.

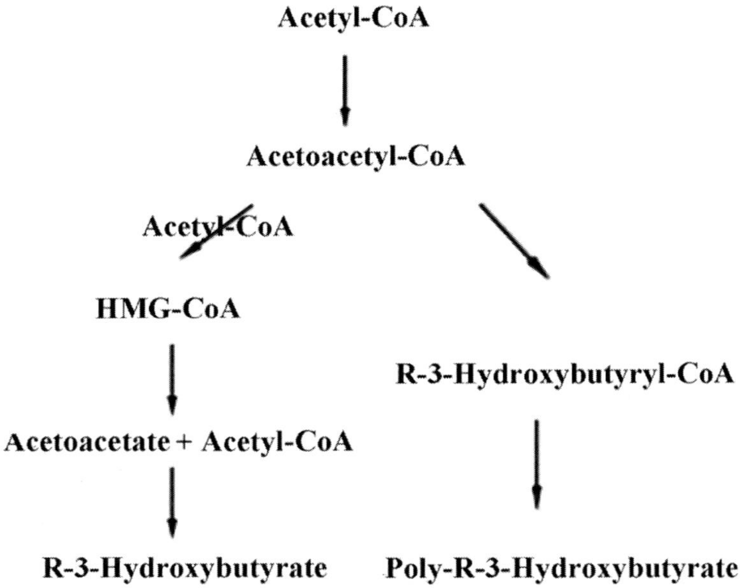

Figure 2. Synthesis of PHB in prokaryotes (right) and R-3-HB in eukaryotes (left). From Reusch 2013.

PHYSICAL PROPERTIES OF CPHB

PHB consisting of six or more residues are insoluble in water and soluble in lipids [9]. The PHB molecule is amphiphilic - hydrophobic methyl groups alternate with hydrophilic carbonyl ester groups (Figure 1); accordingly, the methyl groups may engage in hydrophobic interactions and the ester carbonyl oxygens may serve as hydrogen-bond acceptor or as ligands for coordinate bonds to cations. Moreover, polyesters such as PHB, in contrast to polyamides and polypeptides, have highly flexible backbone structures, since they lack the stabilizing element of internal hydrogen bonds and the rigidity of a peptide bond. The high flexibility of the PHB backbone has been amply demonstrated in studies by Seebach et al., using nuclear magnetic resonance (NMR) spectroscopy [10] circular dichroism and fluorescence (FRET) measurements [11], and molecular dynamics simulations [12]. However, it is important to note that the flexibility of the PHB backbone is highly sensitive to temperature. The glass temperature of PHB is ~10°C [9]; accordingly, the PHB backbone becomes increasingly more rigid as temperatures are lowered below the physiological range.

PHB are members of a class of polymers, known as polymer electrolytes, which have the capacity to 'dissolve' salts [13-15]. Characteristics of this group of polymers are flexible backbones containing electron-donating oxygens that recur at frequent distances so as to allow the polymer chain to form multiple coordinate bonds to cations.

As shown in Figure 3, PHB has the same backbone structure as polypropiolactone (PPL), which forms ion-conducting complexes with lithium perchlorate [13]. The two polyesters differ only in the presence of a methyl group in the beta position of the repeating unit. The effect of this methyl group may be inferred from the parallel situation in the ion-conducting polyethers – PEO and 1,2-PPO – as attenuating but not negating the ability to solvate cations [14, 15].

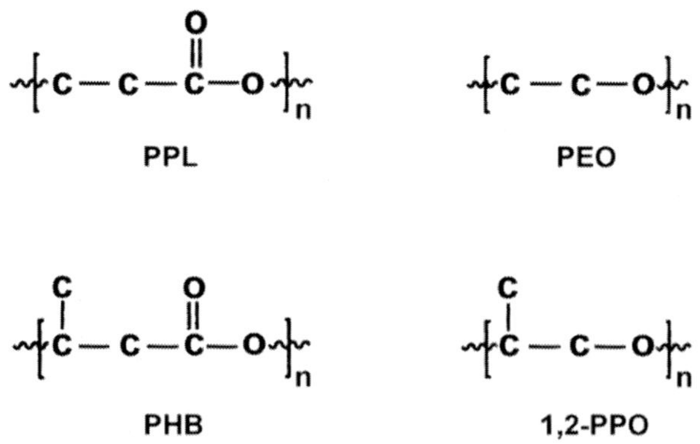

Figure 3. Backbone structure of polymer electrolytes. PEO – polyethylene oxide; 1,2-PPO – 1,2, polypropylene oxide; PPL – polypropiolactone. From Reusch 1992.

The salt-solvating properties of PHB havebeen well-established by crystal structures of salts of cyclic triesters of R-3-hydroxybutanoic acid (triolides) (16), bystudies that demonstrate the ability of short-chain PHB to transport cations across methylene chloride layers in U-tubes (17), of poly-(R)-3-hydroxyalkanoates to form ion-conducting complexes with lithium perchlorate (18), and of mPHB to form nonselective ion channels in planar lipid bilayers (19) and in liposomes (20).

PHBYLATION

In PHBylation, a cPHB molecule forms a covalent bond at its S-CoA ester end to the hydroxy group of an amino acid (usually serine) of a protein, and the methyl groups and ester carbonyl oxygens of the remaining cPHB residues then form multiple noncovalent bonds to neighboring amino acid residues. Individually, these noncovalent interactions between cPHB functional groups and amino acid residues are weak, but the frequency at which the methyl groups and ester carbonyl oxygen atoms repeat along the flexible cPHB backbone allow each polymer molecule to make many such contacts. The number and strength of these bonds would be determined by the primary structure and three-dimensional geometry of the protein near the binding site(s).

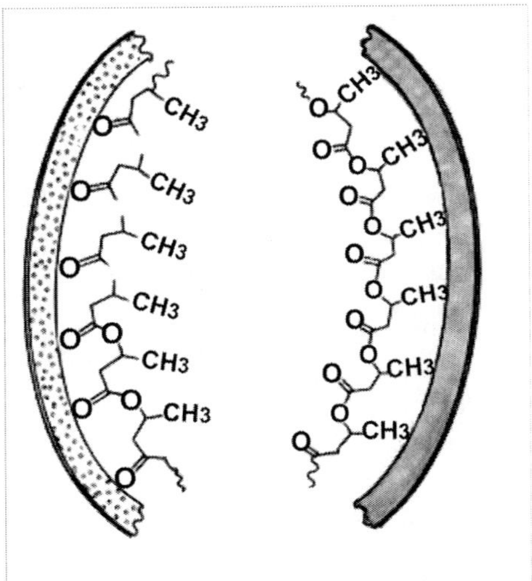

Figure 4. Sketch illustrating binding of cPHB residues to hydrophobic (shaded) and hydrophilic (dotted) regions of polypeptides, effecting a reversal of their polarity.

The solubility of a cPHB-protein is mainly a function of its amino acid composition. However, PHBylation will decrease the aqueous solubility of the peptide and increase its lipid solubility, and thus may influence the location of that peptide within the protein, i.e., transfer it to the inside of a cytoplasmic protein or the outside of a membrane protein. Accordingly, PHBylation may serve to assist the insertion of a hydrophilic peptide into lipid bilayers.

The amphiphilicity and backbone flexibility of cPHB enable them to conform to the contours of peptide segments, and in some cases this may result in adjustment or reversal of the phobicity of peptide segments, i.e., make the surface of a hydrophilic peptide (dotted) more hydrophobic or that of a hydrophobic peptide (shaded) more hydrophilic (Figure 4).

cPHB-Conjugated Proteins

In prokaryotes, cPHB-proteins were first observed in *Escherichia coli* [21] and they have been examined most extensively in this organism [21-23]. They are widely distributed within the *E. coli* cell wherein they are mainly associated with membranes and nucleic acids. cPHB proteins have also been identified in *Haemophilus influenzae* (OprA) [24] and *Streptomyces lividans* (KcsA) [25]. The solvating properties of cPHB and its cellular distribution suggest it may be involved with binding and/or transfer of ions and of polyanions such as inorganic polyphosphates and polynucleotides.

In eukaryotes, cPHB-proteins identified thus far are the CaATPase of human erythrocytes [26] and the transient receptor potential melastatin 8 channel (TRPM8) [27]. The CaATPase was also found to contain inorganic polyphosphate, and it was proposed that the cPHB may be engaged in solvating the calcium polyphosphate salt within the CaATPase pump. TRPM8 is a sensor of environmental temperatures in the peripheral nervous system [28, 29]. The channel is activated by low temperatures and by chemicals such as menthol, icilin and eucalyptol. The temperature range at which the channel is activated (ca 8-26°C) roughly corresponds to the temperature range at which the PHB backbone adjusts from rigid to flexible. Cao et al., [28] identified by mass spectrometry a number of PHB-modified peptides in the N terminus of the TRPM8 protein and in its extracellular linker. They found that removal of cPHB by enzymatic hydrolysis, or by site-directed mutagenesis of the serine residues that serve as covalent anchors for cPHB or the hydrophobic residues adjacent to those serine residues, resulted in significant inhibition of TRPM8 channel activity.

INFLUENCE OF CPHB ON FOLDING OF *E. COLI* OUTER MEMBRANE PROTEIN A (OMPA)

The protein for which the role of PHBylation has been most extensively examined is the outer membrane protein A (OmpA) of *E. coli*, a major protein of the outer membrane and a well-known model for protein folding [30-35]. The many structural and physiological functions of OmpA include maintenance of cell shape and stability, serving as a bacteriophage receptor [36], and acting as a mediator in bacterial conjugation [37].

The OmpA precursor (ProOmpA) is synthesized by ribosomes in the cytoplasm as a 325-residue protein with a 21-residue signal sequence at the amino terminal [30]. The signal sequence is cleaved as the OmpA protein moves to the inner membrane. It crosses the inner membrane, traverses the periplasm, and inserts into the outer membrane, wherein it folds into its mature conformation. During this journey, modifications are made to the protein, first by cytoplasmic enzymes and then by periplasmic enzymes. Studies of the structure of OmpA have been performed using OmpA obtained by over expression of the protein into cytoplasmic inclusion granules (I-OmpA) in *E. coli*, or with OmpA extracted from the outer membranes of *E. coli* (M-OmpA). Under both circumstances, OmpA will have acquired modification(s) that are made to the protein by cytoplasmic enzymes, but only OmpA extracted from the outer membranes will have acquired additional modification(s) to the protein made by periplasmic enzymes.

Modifications to OmpA in the cytoplasm. The OmpA protein consists of an N-terminal domain (residues 1-171) and a C- terminal domain (residues 172-325). Studies by the Henning group have demonstrated that residues 163–170 (SLGVSYRF) of the N-terminal domain are essential for the integration of the protein into the outer membrane [38-41]. When these residues, known as the 'sorting signal', are removed, OmpA remains in the periplasm. Western blot immunoassay and chemical assay studies by Xian et al., [42] revealed that serines 163 and 167 of the sorting signal are modified by cPHB. The two serines are modified by cPHB in OmpA isolated from cytoplasmic inclusion bodies (I-OmpA), and also in OmpA isolated from outer membranes (M-OmpA); accordingly, modifications of the sorting signal by cPHB occur in the cytoplasm.

Xian et al., [42] found that the residues flanking the serines of the sorting signal are critical to their modification by cPHB. PHBylation of the serines occurred only when they were bordered by hydrophobic residues.

Accordingly, when either of the hydrophobic leucine residues adjacent to S163 were replaced by neutral or hydrophilic residues, PHBylation of S163 did not occur, and when the hydrophobic valine residue adjacent to S167 was replaced, PHBylation of S167 did not occur.

cPHB is not discernible in the Raman spectra [43,44], ^{1}H-NMR spectra [45] or X-ray structures [46, 47] of the N-terminal domain of OmpA. Due to its highly flexible backbone, cPHB would either not be visible in these spectra or it would resembles detergent or lipid molecules. However, modification of the peptide containing the sorting signal, i.e., peptide 162–174 (LSLGVSYRFGQGE) (MW 1412) in the eighth β-strand of the narrow pore, by up to ten cPHB residues is detectable in the MALDI-MS spectra [42] (Fig. 5). The actual number of residues may be greater since there may have been some hydrolysis of the ester bonds of cPHB during the isolation and purification of the protein. The MALDI-MS spectra illustrate the lability of the ester bonds of cPHB. In the upper spectrum, the parent peak of the peptide (MW 1412) is barely visible, but strong peaks are observed that correspond to cPHB-modification of this peptide by 6, 8 and 10 R-3HB residues. After ~3 additional seconds of the laser beam (lower spectrum), the peak corresponding to 10 R-3HB has diminished, peaks corresponding to 6 and 8 R-3HB have increased in intensity, and the parent peak of the peptide (MW 1412) is now clearly visible indicating the total removal of cPHB from some peptide molecules. Continued treatment of the peptide with the laser beam further degraded the peaks corresponding to modification by R-3HB with parallel increases in the intensity of the peak corresponding to the parent peptide (not shown).

Planar lipid bilayer studies by Negoda et al., [48] show that PHBylation of serines 163 and 167 of the eighth β-strand is critical to the formation of a well-structured narrow pore by OmpA (Figure 6). The loss of cOHB on serine 167 by mutation of the serine to glycine (S167G) results in the formation of pores with a significantly lower conductance, i.e., narrower pores, while loss of cOHB on serine 163 (S163G) leads to formation of pores with even lower conductance and more frequent closures. The double mutant S163G:S167Gthat is without cOHB on the sorting signal, displays only trial insertions into the bilayer, which indicates that the protein is associated with the bilayer but is unable to form a stable open pore. We may conclude that PHBylation of the eighth beta strand in the cytoplasm is essential for the formation of the narrow pore and essential to the transfer of the OmpA protein into and across the cytoplasmic membrane.

Modifications of OmpA in the periplasm. Further modifications to OmpA occur in the periplasm. Negoda et al., [49] showed by chemical and western blot analysis and by ^{1}H-NMR that the C-terminal segment 264–325 of OmpA is also modified by cPHB. This modification is present in M-OmpA, but not in I-OmpA, hence it must occur in the periplasm. This inference is consistent with the presence of the cPHB polymerase *ydcS* in the periplasm, and with the ~30% decrease in concentration of cPHB in outer membrane proteins in *ydcS* deletion mutants [7]. The specific residue(s) modified by cPHB within this segment have not been identified, and it is also unknown whether other segments of the C-terminal domain are modified by cPHB.

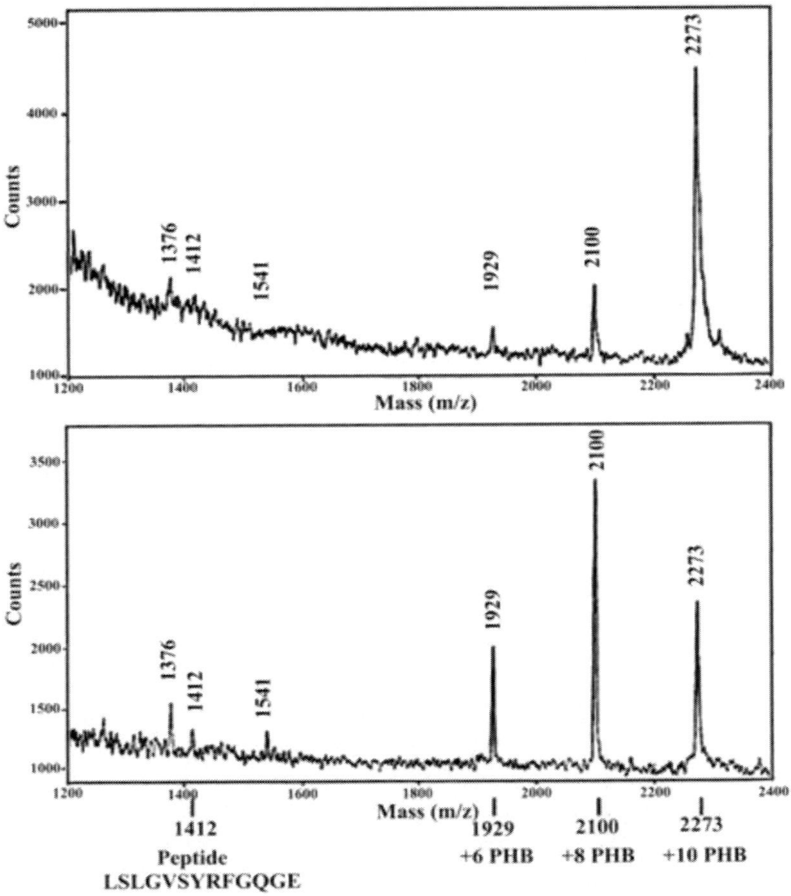

Figure 5. MALDI/MS spectrum of OmpA peptide 162–174 (MW 1412), obtained by cyanogen bromide and *S. aureus* Glu-C digestion of OmpA, showing modification by cPHB. Matrix: α-cyano-4-hydroxycinnamic acid. Upper panel - initial spectrum. Lower panel - spectrum after ~3 additional seconds of laser beam. From Xian et al., 2007.

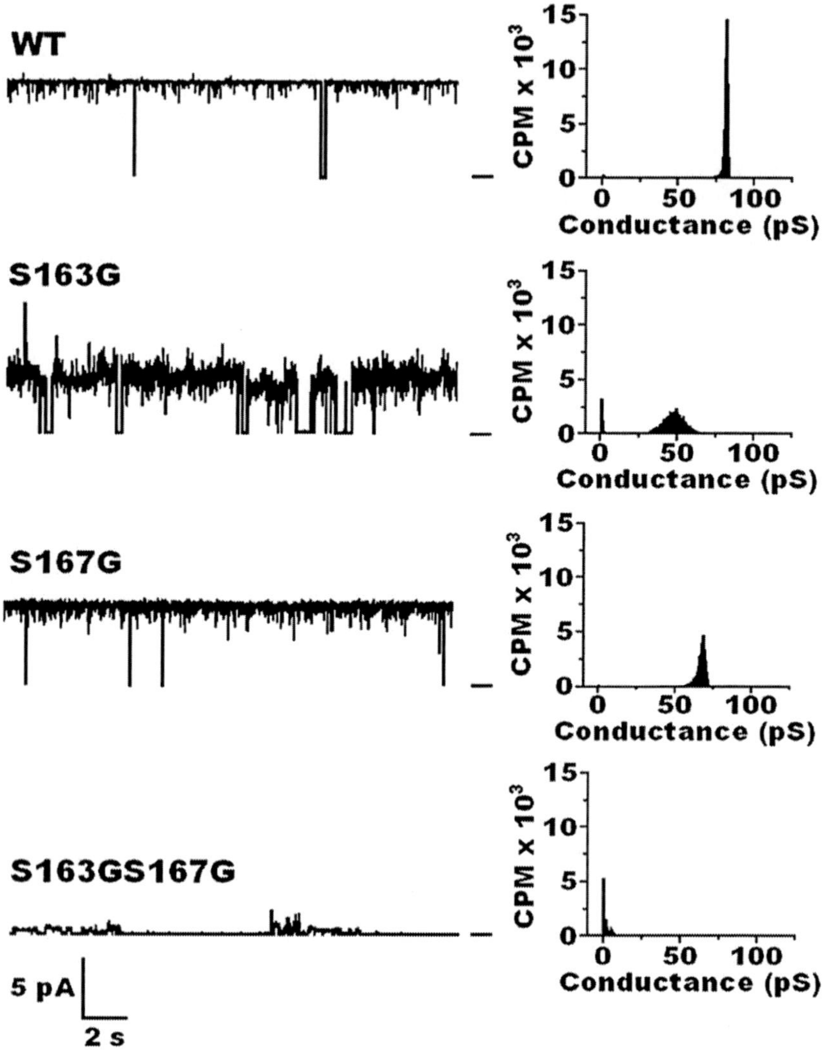

Figure 6. Left: Representative current traces of OmpA wild-type and mutant proteins isolated from cytoplasmic inclusion bodies of *E. coli* BL21 (DE3) pLysS cells, reconstituted into C8E4 micelles, and incorporated into planar lipid bilayers of DPhPC (diphytanoylphosphatidylcholine) at room temperature between aqueous solutions of 1 M KCl in 20 mM HEPES, pH 7.4. Bar at right of the traces indicates the closed state of the channel. Clamping potential was +100 mV. Right: Histograms taken from records of ~10 minutes at room temperature. From Negoda et al., 2010, BBA.

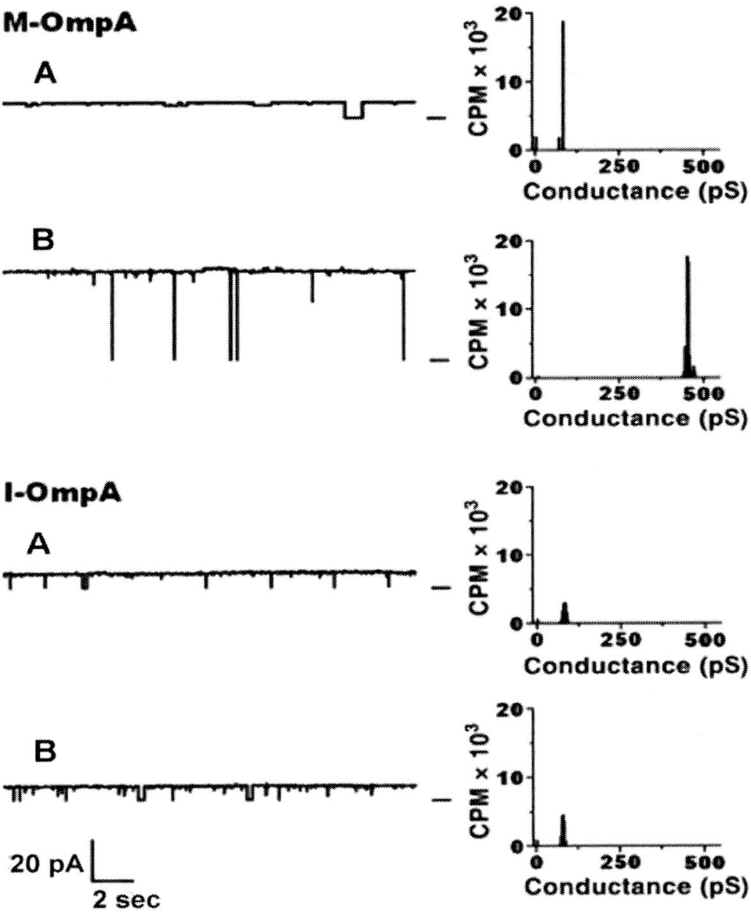

Figure 7. Representative single-channel current traces of M-OmpA and I-OmpA in lipid bilayers of DPhPC between aqueous solutions of 20 mM Hepes (pH 7.4) and 1 M KCl at 22°C. Upper traces (A): M-OmpA and I-OmpA at 22°C. Lower traces (B): M-OmpA and I-OmpA at 22°C after incubation at 40°C for 2 h. The closed state is indicated by the bar at the right of each trace. The clamping potential was +100 mV with respect to ground (trans). The corresponding histograms from 1 min of continuous recording show the distribution of conductance magnitudes. CPM, counts per minute. From Negoda et al., 2010, FEBS Lets.

Planar lipid bilayer studies by Zakharian and Reusch [50, 51] and by Negoda et al., [49] show that both M-OmpA and I-OmpA form narrow pores with a high open time at room temperatures (Figure 7A). However, after incubation at elevated temperatures, M-OmpA undergoes a transition into large pores (~450 pS) (Figure 7B), whereas I-OmpA remains a narrow pore. This refolding of OmpA from a narrow to a large pore is strongly temperature-dependent. The transition of a single molecule of OmpA in a planar lipid bilayer required ~ 2 days at 26°C, ~ 2 hours at 30°C, ~ 30 minutes at 37°C and ~ 10 minutes at 42°C (Ea = 33.2 kcal/mol). *In vivo*, the activation energy may be lowered by chaperones that assist the folding process.

Negoda et al., [49] identified yet another modification that occurs in the periplasm as the formation of a disulfide bond between the two cysteines - C290 and C302 - in the C-terminal domain of OmpA by the enzyme DsbA [52, 53]. When the disulfide bond of the narrow pore conformer is broken by addition of reducing agents, such as mercaptoethanol or dithiothreitol, the heat-induced transition to the large pore conformation does not occur. However, once the transition to the large pore conformer has been completed, the addition of disulfide-bond reducing agents has no effect, i.e., OmpA remains a large pore after losing its disulfide bond. Accordingly, an intact disulfide bond is essential to the formation of the large pore conformation, but necessary for maintaining the large pore conformation.

We may conclude that OmpA cannot fold into its mature structure without modifications by cPHB of its N-terminal domain in the cytoplasm and its C-terminal domain in the periplasm. It cannot fold into its mature structure without formation of a disulfide bond in the C-terminal domain in the periplasm. Even with all of these modifications, it does not fold into its mature structure in a reasonable time frame below physiological temperatures. However, once the rearrangement from narrow to large pore has been completed, the large pore conformation is very stable. Extended periods at low temperatures do not reverse the transition. Breaking the disulfide bond does not reverse the transition. Large pores are converted to narrow pores only when they are subjected to denaturing conditions such as the addition of ionic detergents. A large pore is the final product of the folding cascade of OmpA. The narrow pore conformer is a folding intermediate in the living *E. coli* cell; not a normal structure in its outer membrane. The large pore structure is consistent with the known functions of OmpA as a receptor for bacteriophages [36] and participant in bacterial conjugation [37], both of which indicate it forms a pore large enough for the passage of ssDNA.

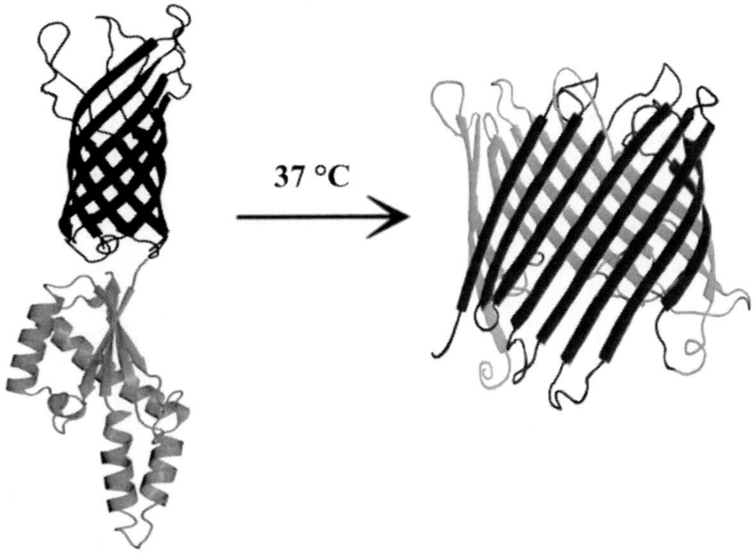

Figure 8. Transition of narrow pore conformation of OmpA to large pore conformation following modifications by cPHB, disulfide-bond formation.

CONCLUSION

cPHB are a class of protein conjugates with important roles in protein folding. The physical properties of cPHB suggest they also participate in the targeting and functions of these proteins. While protein over expression is a convenient method of obtaining ample amounts of protein for physical studies, a protein is more than a chain of amino acids. The role(s) of protein modifiers are critical to the folding, structure and function(s) of the mature protein, and not all modifications take place in the cytoplasm. It is also important to realize that although the polypeptide structure is not strongly temperature-dependent, modifiers such as lipids and cPHB may make it so. Unless all temperature-sensitive modifications have been ruled out, protein folding should always be observed within the temperature range at which it occurs in nature.

REFERENCES

[1] Anderson AJ; Dawes EA. Occurrence, metabolism, metabolic role and industrial uses of bacterial polyhydroxyalkanoates. *Microbiol. Rev.,* 1990, 54, 450-472.

[2] Jendrossek D. Polyhydroxyalkanoate granules are complex subcellular organelles (carbonosomes). *J. Bacteriol.,* 2009, 191, 3195-3202.

[3] Reusch RN. Biological complexes of poly-beta-hydroxybutyrate. *FEMS Microbiol. Rev.*, 1992, 9, 119-129.

[4] Reusch RN. Nonstorage poly-(R)-3-hydroxyalkanoates (complexes PHAs) in prokaryotes and eukaryotes. 2005 *Biopolymers Online.*

[5] Reusch RN. Physiological importance of poly-(R)-3-hydroxybutyrates. *Chem. & Biodiver.* 2012, 9, 2343-2366.

[6] Wahab HA; Khairudin MR; Samian MR; Najimudin N. Sequence analysis and structure prediction of type II *Pseudomonas sp.* USM 4-55 PHA synthase and an insight into its catalytic mechanism, *BMC Struct. Biol.* 2006, 6, 6–23.

[7] Dai D.; Reusch RN. Poly-3-hydroxybutyrate synthase from the periplasm of *Escherichiacoli. Biochem. Biophys. Res. Comm.* 2008, 374, 484-489.

[8] Lehninger AL. Biochemistry; *Worth Publishers Inc.:* New York, NY, USA, 1970; p. 427.

[9] Holmes PA. In Developments in crystalline polymers. 1988. Bassett DC (ed). *Elseview Applied Sci.* London.

[10] Albert M; Seebach D. Synthesis and NMR analysis in solution of oligo (3-hydroxyalkanoic acid). Derivatives with the side chains of alanine, valine, and leucine (β- depsides): coming full circle from PHB to β-peptides to PHB. *Helv. Chim. Acta* 2002, 85, 633-655.

[11] Rueping M; Dietrich A; Buschmann V; Fritz MG; Sauer M; Seebach D. On the structure of poly(3-hydroxybutanoic acid) in solution and in phospholipid bilayers. Circular dichroism and fluorescence spectroscopy with oligo (3-hydroxybutanoic acid) derivatives. *Macromol.* 2001, 34, 7042-7048.

[12] Gee PJ; Hamprecht FA; Schuler LD; van Gunsteren WF; Duchardt E; Schwalbe H. Molecular-dynamics simulation study of the conformational preferences of oligo(3-hydroxyalkanoic acids) in chloroform solution. *Helv. Chim. Acta,* 2002, 85, 618-632.

[13] Watanabe M; Togo M; Sanui K; Ogatu N; Kobayashi T; Ohtaki, Z. Ionic conductivity of polymer complexes formed by poly(β-propiolactone) and lithium perchlorate, *Macromol.* 1984, 17, 2908–2912.

[14] MacCallum J R; Vincent CA. Ion–molecule and ion–ion interactions in polymer electrolytes. In: *Polymer Electrolyte Reviews* 1, 1987, 23-37.

[15] Armand MB. Current state of PEO-based electrolyte, In: *Polymer Electrolyte Reviews* 1, 1987, 1-22.

[16] Seebach D; Bürger M; Plattner DA; Nesper R; Fässler T. Cylic oligomers of R-3- hydroxybutanoic acid: Preparation and structural aspects. *Helv. Chim. Acta* 1993, 76, 2581- 2601.

[17] Bürger HM; Seebach D. Cation transport across bulk liquid organic membranes with oligomers of (R)-3-hydroxybutanoic acid. *Helv. Chim. Acta* 1993, 76, 2570-2580.

[18] Reusch RN; Reusch WH. Branched polyhydroxyalkanoate polymer salt electrolytic compositions and method of preparation. 1996, U.S. Patent No. 5, 266, 422.

[19] Seebach D; Brunner A; Bürger HM; Reusch RN; Bramble LL. Channel-forming activity of 3-hydroxybutanoic-acid oligomers in planar lipid bilayers. *Helv. Chim. Acta* 1996, 79, 507-517.

[20] Fritz MG; Walde P; Seebach D. Detection, synthesis, structure and function of oligo(3-hydroxyalkanoates): contributions by synthetic organic chemists. *Macromolecules* 1999, 32, 574-580.

[21] Huang R; Reusch RN. Poly(3-hydroxybutyrate) is associated with specific proteins in the cytoplasm and membranes of *Escherichia coli*. *J. Biol. Chem.* 1996, 271, 22196-202.

[22] Reusch RN; Grühn AG. Poly(3)-hydroxybutyrate-conjugated proteins are ubiquitous inboth prokaryotes and eukaryotes. *Int. Symp. Bacterial.* Steinbüchel, A., Poirier, Y., Witholt, B., Eds.), Ottawa: *NRC Res. Press,* 1997, 10–19.

[23] Reusch RN; Shabalin O.; Crumbaugh A; Wagner R; Schröder; Wurm R. Posttranslational modification of *E. coli* histone-like protein H-NS and bovine histones by short-chain poly- (R)-3-hydroxybutyrate. *FEBS Letts.* 2002, 527, 319-322.

[24] Zakharian E; Reusch RN. *Haemophilus influenzae* outer membrane protein P5 is associated with inorganic polyphosphate and polyhydroxy-butyrate. *Biophys. J.* 2007, 92,588-593.

[25] Reusch RN. *Streptomyces lividans* potassium channel contains poly-(R)-3-hydroxybutyrateand inorganic polyphosphate. *Biochem.* 1999, 38, 15666-15672.

[26] Reusch RN; Huang R; Kosk-Kosicka D. Novel components and enzymatic activities of the human erythrocyte plasma membrane calcium pump. *FEBS Letts*. 1997, 412, 592-596.

[27] Zakharian E; Cao C; Rohacs T. Gating of transient receptor potential melastatin 8 (TRPM8) channels activated by cold and chemical agonists in planar lipid bilayers. *J. Neurosci.* 2010, 15, 12526-12534.

[28] Cao C; Yudin Y; Bikard Y; Chen W; Liu T; Li H; Jendrossek D; Cohen A; Pavlov E; Rohacs T; Zakharian E. Polyester modification of the mammalian TRPM8 channel protein: implications for channel structure and function. *Cell Rep*. 2013, 4, 302-315.

[29] Bautista DM; Siemens J; Glazer JM; Tsuruda PR; Basbaum AI; Stucky CL; Jordt S-E; Julius D. The menthol receptor TRPM8 is the principal detector of environmental cold. *Nature*. 2007, 448, 204–208.

[30] Surrey T.; Jähnig F. Kinetics of folding and membrane-insertion of a β-barrel membrane protein. *J. Biol. Chem*. 1995, 270, 28199-28203.

[31] Kleinschmidt JH; Tamm LK. Time-resolved distance determination by tryptophan fluorescence quenching: probing intermediates in membrane protein folding. *Biochem.* 1999, 38, 4996-5005.

[32] Kleinschmidt JH; den Blaauwen T; Driessen AJ; Tamm LK. Outer membrane protein A of *Escherichia coli* inserts and folds into lipid bilayers by a concerted mechanism. *Biochem.*1999, 38, 5006-5016.

[33] Tamm LK; Arora A; Kleinschmidt JH. Structure and assembly of β-barrel membrane proteins. *J. Biol. Chem.* 2001, 276, 32399-32402.

[34] Anderson KK; Otzen DE. Folding of outer membrane protein A in the anionic biosurfactant rhamnolipid. *FEBS Let.* 2014, 588, 1955-1960.

[35] Koebnik R; Locher KP; Van Gelder P. Structure and function of bacterial outer membrane proteins: barrels in a nutshell. *Mol. Microbiol.* 2000, 37, 239–253.

[36] Morona R; Klose M; Henning U. *Escherichia coli* K-12 outer membrane protein (OmpA) as a bacteriophage receptor: analysis of mutant genes expressing altered proteins. *J. Bacteriol.* 1984, 159, 570–578.

[37] Schweizer M; Henning U. Action of major outer cell envelope membrane protein in conjugation of *Escherichia coli* K-12. *J. Bacteriol.* 1977, 129, 1651–1652.

[38] Bremer E; Cole ST; Hindemach I; Henning U; Beck E; Kurz C; Schaller H. Export of a protein into the outer membrane of *Escherichia coli* K12. Stable incorporation of the OmpA protein requires less than193 amino-terminal amino-acid residues. *Eur. J. Biochem.*1982, 122, 223–231.

[39] Klose M; Schwarz H; MacIntyre S; Freudl R; Eschbach M; Henning U. Internal deletions in the gene for an *Escherichia coli* outer membrane protein define an area possibly important for recognition of the outer membrane by this polypeptide. *J. Biol. Chem.* 1988, 263, 13291-13296.

[40] Klose M; MacIntyre S; Schwarz H; Henning U. The influence of amino substitutions within the mature part of an *Escherichia coli* outer membrane protein (OmpA) on assembly of the polypeptide into its membrane, *J. Biol. Chem.* 1988, 2634, 13297-13302.

[41] Freudl R; Klose M; Henning U. Export and sorting of the *Escherichia coli* outer membrane protein OmpA. *J. Bioenerg. Biomembr.* 1990, 22, 441-449.

[42] Xian M; Fuerst MM; Shabalin Y; Reusch RN. Sorting signal of *Escherichia coli* OmpA is modified by oligo-(R)-3-hydroxybutyrate. *Biochim Biophys Acta* 2007, 1768, 2660-2666.

[43] Vogel H; Jähnig F. Models for the structure of outer-membrane proteins of *Escherichia coli* derived from raman spectroscopy and prediction methods. *J. Mol. Biol.* 1986, 190,191-199.

[44] Sanchez KM; Neary TJ; Kim JE. Ultraviolet resonance raman spectroscopy of folded and unfolded states of an integral membrane protein. *J. Phys. Chem.* 2008, 112, 9507-9511.

[45] Zoonens M.; Catoire L. J.; Giusti F.; Popot J. L. NMR study of a membrane protein in detergent-free aqueous solution. *Proc. Natl. Acad. Sci USA* 2005, 102, 8893-8898.

[46] Pautsch A; Schulz GE. Structure of the outer membrane protein A transmembrane domain, *Nat. Struct. Biol.* 1998, 5, 1013–1017.

[47] Pautsch A; Schulz GE. High-resolution structure of the OmpA membrane domain, *J. Mol. Biol.* 2000, 298, 273–282.

[48] Negoda A; Negoda E; Reusch RN. Oligo-(R)-3-hydroxybutyrate modification of sorting signal enables pore formation by *Escherichia coli* OmpA. *Biochim. Biophys. Acta* 2010, 1798, 1480-1484.

[49] Negoda A; Negoda E; Reusch RN. Resolving the native conformation of *Escherichia coli* OmpA. *FEBS J.* 2010, 277, 4427-4437.

[50] Zakharian E; Reusch RN. Outer membrane protein A of *Escherichia coli* forms temperature-sensitive channels in planar lipid bilayers. *FEBS Lett.* 2003, 555, 229-235.

[51] Zakharian E; Reusch RN. Kinetics of folding of *Escherichia coli* OmpA from narrow to large pore conformation in a planar bilayer. *Biochem.* 2005, 44, 6701-6707.

[52] Bardwell JC; McGovern K; Beckwith J. Identification of a protein required for disulfide bond formation *in vivo*. *Cell* 1991, 67, 581–589.
[53] Ito K; Inaba K. The disulfide bond formation Dsb system. *Curr. Opin. Struct. Biol.* 2008, 18, 450–458.

In: Advances in Protein Folding Research ISBN: 978-1-63483-593-0
Editor: Mathew Hale

Chapter 2

Calculation of Free-Energy Profiles, Folding Rates and Φ Values by Means of a Simple Statistical-Mechanical Model of Protein Folding

Hiroshi Wako[1,*] and Haruo Abe[2]
[1]School of Social Sciences, Waseda University, Tokyo, Japan
[2]Nishinippon Institute of Technology,
Fukuoka, Japan

Abstract

Analysis of Φ values provides information on structures of transition states along a folding pathway of a protein via measurement of effects of a mutation on folding kinetics. In this chapter, we discuss the Φ values calculated for 27 proteins (54–128 amino acid residues long) by means of a simple statistical-mechanical model of protein folding (developed by us); we compare these data with experimentally observed Φ values. In the model, three-dimensional structure of a protein was taken into consideration via inter-residue contact information of the native structure obtained from the Protein Data Bank. A free-energy profile of the wild-type protein was calculated by assigning a single value to interaction energies of the inter-residue contacts (regardless of the amino acid types) as an approximation.

* E-mail address: wako@waseda.jp; Address: School of Social Sciences, Waseda University, Tokyo 169-8050, Japan (Corresponding author)

The same procedure was applied to a protein with a single point mutation but with slightly greater or smaller values assigned to the interaction energies between the mutated and other (unmutated) residues. The folding and unfolding rates were individually estimated from the free-energy profiles of the wild-type protein and mutant protein. The Φ value was then calculated for every residue from the differences in the folding or unfolding rates between the wild-type and the mutant. Despite the simplicity of the model, corresponding experimental Φ values were fairly well predicted for most proteins. The successful cases point to the importance of topology of the three-dimensional structure and support validity of assumptions of the model. Conversely, the cases of failure are suggestive of the need for interaction details such as amino acid types and interaction modes during the calculation and revealed limitations of the model. Protein-folding mechanisms, namely the framework model and nucleation-condensation mechanism, are also discussed in terms of the advantages and disadvantages of the proposed model.

1. INTRODUCTION

1.1. The Protein-Folding Problem

Genetic information is encoded in the nucleotide sequence of DNA. The DNA sequences are translated into amino acid sequences of proteins. Every protein has a unique amino acid sequence and consequently its own unique three-dimensional (3D) structure determined by the amino acid sequence. One or more proteins are usually participating in a life-related function, and this function is strongly linked with the 3D structure. A living organism is a highly organized system of interacting biomolecules including proteins. In this scheme of flow of genetic information (according to the Central Dogma of molecular biology), the sequence-structure and structure-function relations of proteins have been challenging problems in life sciences.

In particular, the protein-folding problem is one of such challenging topics, especially interesting from biochemical and biophysical perspectives. The protein-folding problem deals mainly with three questions: (1) how 3D structure of a protein is determined by its amino acid sequence, (2) how a protein folds from a random coil into native structure, and (3) what the static and dynamic properties of the 3D structure are. Question (2) is the focus of this chapter.

Various scientific principles have been proposed to explain the mechanism of protein folding from a random coil into native structure. Regarding the question of how a protein molecule can find its native structure spontaneously, the following two principles are widely cited. One is the consistency principle of Gō (1983), which stipulates that various energetic factors that stabilize the 3D structure of a protein are evolutionarily designed to be consistent with one another. Another is the minimal-frustration principle, which was proposed by Bryngelson and Wolynes (1987). This principle states that the energetic frustration exists in the native structure, in other words, energetic factors are not perfectly consistent, but this frustration is minimized. As for the question of how a protein can fold so rapidly—overcoming Levinthal's paradox (Levinthal 1968)—the funnel hypothesis, i.e., a funnel-shaped energy landscape with a downhill slope toward the native structure, is now a common view (Bryngelson et al. 1995; Dill and Chan 1997). Although these principles differ in their expression, they essentially emphasize specificity of the native structure, in comparison with the other (non-native) conformations.

The folding process (or folding pathway) is another problem. There are two major hypotheses: one is the three-step mechanism or framework model (Anfinsen 1972; Tanaka and Scheraga 1977; Kim and Baldwin 1982; Ptitsyn 1987), and the other is the nucleation-condensation model (Itzhaki et al. 1995; Fersht 1997). In the framework model, secondary-structure elements fold first, followed by coalescence of the preformed secondary-structure elements into the native structure. Step-by-step formation from short- to medium- to long-range interactions is the key assumption. Although some researchers do not consider the step-by-step formation as an essential requirement of the framework model, this condition is imposed in the present chapter to make the assumptions of our model clear and to accentuate differences from the nucleation-condensation model. On the other hand, in the nucleation-condensation model, secondary and tertiary structures are believed to form in parallel. In this model, the long-range interactions can be formed before or simultaneously with formation of the secondary structure.

The above-mentioned key principles have contributed significantly to this field of research and helped to provide a conceptual basis for protein folding. Nonetheless, there is a critical limitation: the difficulty in connecting the theory to experimental results quantitatively (Muñoz 2014). Experimentalists could not use the existing theory to interpret their results. Such limitations have been progressively overcome by means of improved simulations (including molecular dynamics) and analytical calculations with simple statistical-mechanical models. Because computer simulations are a time-consuming task and can usually be performed only for one protein at a time, there is a need for simple statistical-mechanical models of protein folding, which allow for direct analysis and fitting of experimental data.

An Ising-like statistical-mechanical model of protein folding was developed for the first time by Wako and Saitô (1978a, 1978b). This model was elaborated by Gō and Abe (1981), who demonstrated that it could well reproduce a simulation of two-dimensional (2D) lattice proteins (Abe and Gō 1981). Nevertheless, it took a long time for a similar model to be applied to real proteins by Muñoz and Eaton (1999), who showed that theoretically calculated folding rates correlate well with experimentally observed ones.

In this chapter, we discuss the Φ values that are calculated for every residue of a given protein with a simple statistical-mechanical model; we also compare these data with the corresponding experimental Φ values. Despite the simplicity of the model, the experimental Φ values were plausibly reproduced for most proteins. Although the successful cases support the validity of the assumptions of the model, the cases of failure revealed limitations of the model under such assumptions. We will discuss protein-folding mechanisms while examining these successful cases and failures.

1.2. Φ Value Analysis

The Φ value analysis was introduced by Fersht and coworkers (Matouschek et al. 1989; Fersht et al. 1992) to characterize the transition-state structures along the folding pathway of a protein by measuring the effects of mutations on folding kinetics. The Φ value is calculated as $\Delta\Delta G_{\ddagger\text{-}D}/\Delta\Delta G_{N\text{-}D}$ for a two-state-folding protein, where $\Delta G_{\ddagger\text{-}D}$ and $\Delta G_{N\text{-}D}$ are the free-energy differences between an unfolded state and transition state and between an unfolded and folded state, respectively, and ΔΔ denotes the change in these free-energy differences brought about by a point mutation in some amino acid residue. A Φ value of 1.0 indicates that all of the interactions with the mutated residue are formed in the transition state, whereas a Φ value of zero means that the residue is not involved in stabilization of the transition state. Intermediate Φ values indicate that the interactions are formed partially or that there are two populations of mostly unfolded and mostly folded states. On the other hand, the relation between the actual Φ value and the extent of structure formation is not necessarily linear; thus, the interpretation of the intermediate Φ values is still controversial (Fersht and Sato 2004).

The Φ values have been studied experimentally for many two-state-folding proteins (Table 1). The Φ values obtained in these experiments have provided valuable information on protein-folding mechanisms such as the nucleation-condensation mechanism and framework model (Nölting and Agard 2008).

Furthermore, the Φ values have been studied from a theoretical point of view. Because folding simulations can provide information on the folding pathway, the transition-state structures in the simulations are associated with experimentally observed Φ values (Alonso and Daggett 2000; Settanni et al. 2005; Banachewicz et al. 2011b). More directly, the Φ values were calculated on the basis of the free-energy profiles obtained by means of a statistical-mechanical model of protein folding (Portman 1998; Galzitskaya and Finkelstein 1999; Alm and Baker 1999; Muñoz and Eaton 1999; Alm et al. 2002; Zong 2006; Kubelka et al. 2008). Although many models have been proposed, their essential concepts of the modeling of protein folding are similar: each residue is assumed to be in one of two states: a native (or folded) state and a random-coil (or unfolded) state. One or more contiguous segments consisting of residues in the native state are assumed to form in a polypeptide chain, and only the interactions between the residues in contact in the native structure are considered (the so-called Gō model); no interactions are assumed to occur in segments consisting of residues in the random-coil state. The calculated Φ values were generally consistent with the experimental Φ values.

Table 1. Proteins used in this study[a),f)]

# Protein name[b)]	PDB code[c)]	No. of residues[d)]	Folding type[e)]
1 engrailed homeodomain	1ENH	54	All α
2 cMyb-transforming protein	1IDY	54	All α
3 B1 IgG-binding domain of protein G	1PGB	56	α+β
4 Src SH3 domain	1SRM	56	All β
5 α-spectrin SH3 domain	1SHG	57	All β
6 Fyn SH3 domain	1FYN(84–142)	59	All β
7 B domain of protein A	1SS1	62	All α
8 DNA binding protein Sso7d	1BF4	63	All β
9 chymotrypsin inhibitor 2	3CI2	64	α+β
10 B1 IgG-binding domain of protein L	2PTL(15–78)	64	α+β
11 cold-shock protein	1CSP	67	All β
12 ubiquitin	1UBQ	76	α+β
13 procarboxypeptidase A2 active domain	1AYE(4A–83A)	78	α+β
14 acyl-coenzyme A binding protein	2ABD	86	All α

Table 1. Continued

# Protein name[b)]	PDB code[c)]	No. of residues[d)]	Folding type[e)]
15 TI I27 domain	1TIU	89	All β
16 barstar	1BTB	89	α/β
17 TNfn3 domain (tenascin)	1TEN	90	All β
18 FNfn10 domain (fibronectin)	1TTF	94	All β
19 U1A	1URN	96	α+β
20 ribosomal protein L23	1N88	96	α+β
21 ribosomal protein S6	1RIS	97	α+β
22 acylphosphatase	2ACY	98	α+β
23 FKBP12	1FKB	107	α+β
24 barnase	1RNB	109	α+β
25 villin 14T	2VIL	126	α+β
26 azurin	1AZU	126	All β
27 CheY	3CHY	128	α/β
28 villin thermostable subdomain	1VII	36	All α
29 formin binding protein	1E0L	37	All β
30 WW prototype	1E0M	37	All β
31 PSB domain of DLAT	2PDD	43	All α
32 albumin-binding domain	1PRB(7–53)	47	All α
33 E3-binding domain of BBL	2WXC	47	All α
34 LysM domain of MltD	1E0G	48	α+β
35 peripheral-subunit binding domain	1W4J	51	All α
36 hTRF1	1BA5	53	All α
37 ribosomal protein L9	1DIV(1–56)	56	α+β
38 Pit1 homeodomain	1AU7(103–160)	58	All α
39 actin-binding protein 1 SH3 domain	1JO8	58	All β
40 Myb domain of human RAP1	1FEX	59	All α
41 434 Cro protein	2CRO	65	All α
42 cold-shock protein (B. caldolyticus)	1C9O	66	All β
43 cold-shock protein (T. maritima)	1G6P	66	All β
44 hypothetical protein EC0298	1RYK	69	All α
45 cold-shock protein, CspA	1MJC	69	All β
46 FF domain	1UZC	69	All α
47 de novo designed 3-helix bundle	2A3D	73	All α
48 Ras-binding domain of c-raf-1	1RFA	78	α+β
49 cytochrome b5	1EHB	82	α+β
50 C-domain of spore coat protein S	1PRS (91–173)	83	All β
51 His-containing phosphocarrier protein	1POH	85	α+β
52 colicin immunity protein Im7	1AYI	86	All α

# Protein name[b)]	PDB code[c)]	No. of residues[d)]	Folding type[e)]
53 colicin immunity protein Im9	1IMQ	86	All α
54 acyl-CoA-binding protein	1ST7	86	All α
55 chitinase A1 fn III domain	1K85 (559–644)	86	All β
56 PI3-kinase SH3 domain	1PNJ	86	All β
57 HypF N-terminal domain	1GXT	88	α+β
58 protein S	1PRS (1–90)	90	All β
59 ribosomal protein L9	1DIV (58–149)	92	α+β
60 twitchin Ig 18'	1WIT	93	All β
61 Suc1	1SCE	97	α+β
62 acylphosphatase	1APS	98	α+β
63 Src tyrosine kinase SH2 domain	1SPR	103	α+β
64 cytochrome b_{562}	256B	106	All α
65 α spectrin, repeat 16	1CUN (7–112)	106	All α
66 α spectrin, repeat 17	1CUN (113–219)	107	All α
67 p13MTCP1	1QTU (1–109)	109	All β
68 myotrophin	2MYO	118	α+β
69 hisactophilin	1HCD	118	All β
70 ileal lipid binding protein	1EAL	127	All β
71 cellular retinol-binding protein II	1OPA	133	All β
72 cellular retinoic acid-binding protein I	1CBI	136	All β

a) This table is mainly based on data from Garbuzynskiy et al., (2013). Data on Φ value analysis are available for proteins ## 1–27. Data on the folding rate are available for proteins ## 1–72 except for #26 (azurin).

b) For some proteins, an abbreviated name is used for convenience.

c) The chain region is indicated in parentheses if the kinetic study and/or the calculation uses only a part of the PDB data. Some PDB data that we used in this study are different from the data from Garbuzynskiy et al., (2013).

d) The number of residues for calculation of a Φ value and/or folding rate is shown. In some cases, the number of residues is different between the experiment and calculation because some residues are missing in the PDB data.

e) The folding type is defined in reference to SCOP database (Murzin et al., 1995).

f) The literary references regarding the analysis of experimental Φ values and/or midtransition folding and unfolding rates are as follows: #1 and #2 Gianni, et al., (2003); #3 McCallister et al., (2000), Maxwell et al., (2005); #4 Riddle et al., (1999), Maxwell et al., (2005); #5 Martinez and Serrano (1999), Maxwell et al., (2005); #6 Northey et al., (2002), Maxwell et al., (2005); #7 Sato et al., (2004), Myers and Oas (2001); #8 Guerois and Serrano (2000); #9 Itzhaki et al., (1995), Maxwell et al., (2005); #10 Kim et al., (2000), Maxwell et al., (2005); #11 Garcia-Mira et al., (2004), Perl et al., (1998); #12 Went and Jackson (2005), Maxwell et al., (2005); #13 Villegas et al., (1998), Maxwell et al., (2005); #14 Kragelund et al., (1999), Maxwell et al., (2005); #15 Fowler and Clarke (2001); #16 Nölting and Andert (2000), Schreiber and

Fersht (1993); #17 Hamill et al., (2000), Clarke et al., (1997); #18 Cota et al., (2001), Cota and Clarke (2000); #19 Ternström et al., (1999), Maxwell et al., (2005); #20 Hedberg and Oliveberg (2004), Maxwell et al., (2005); #21 Otzen and Oliveberg (2002), Maxwell et al., (2005); #22 Chiti et al., (1999), Taddei et al., (1999); #23 Fulton et al., (1999), Maxwell et al., (2005); #24 Serrano et al., (1992), Matouschek et al., (1990); #25 Choe et al., (1998, 2000); #26 Wilson and Wittung-Stafshede (2005); #27 López-Hernández and Serrano (1996), Muñoz et al., (1994); #28 Wang et al., (2003); #29 and #30 Ferguson et al., (2001); #31 Spector and Raleigh (1999); #32 Wang et al., (2004); #33 Neuweiler et al., (2009); #34 Nickson et al., (2008); #35 Sharpe et al., (2008); #36 Gianni, et al., (2003); #37 Maxwell et al., (2005); #38 Banachewicz et al., (2011a); #39 Maxwell et al., (2005); #40 Gianni, et al., (2003); #41 Laurents et al., (2000); #42 and #43 Perl et al., (1998); #44 Maxwell et al., (2005); #45 Reid et al., (1998); #46 Jemth et al., (2005); #47 Zhu et al., (2003); #48 Maxwell et al., (2005); #49 Manyusa and Whitford (1999); #50 Wenk et al., (1998); #51 van Nuland et al., (1998); #52 and #53 Maxwell et al., (2005); #54 Teilum et al., (2005); #55 Lappalainen et al., (2008); #56 Guijarro et al., (1998); #57 Calloni et al., (2003); #58 Wenk et al., (1998); #59 Maxwell et al., (2005); #60 Clarke et al., (1999); #61 Schymkowitz et al., (2000); #62 and #63 Maxwell et al., (2005); #64 Wittung-Stafshede et al., (1999); #65 and #66 Scott et al., (2004); #67 Roumestand et al., (2001); #68 Lowe and Itzhaki (2007); #69 Liu et al., (2002); #70 Dalessio and Ropson (2000); #71 and #72 Burns et al., (1998).

In this chapter, we report the results of calculation of Φ values using a previously developed alternative and simple statistical-mechanical model of protein folding (Abe and Wako 2009; Wako and Abe 2011). Although this model was developed earlier than the above-mentioned models (Wako and Saitô 1978a, b; Abe and Gō 1981; Gō and Abe 1981), various researchers recently adopted or extended this model to study the protein-folding problem (Itoh and Sasai 2004; Imparato et al. 2007; Itoh and Sasai 2010; Bruscolini and Naganathan 2011; Caraglio and Pelizzola 2012; Sivanandan and Naganathan 2013; Inanami et al. 2014). We also determined the Φ values for lattice proteins using this alternative model, with a particular focus on the dependence of the properties of lattice proteins on the amino acid sequence and the folding topology (Wako and Abe 2011). In line with these studies, we calculated the Φ values of real proteins. For the calculations, we selected 27 proteins (54–128 residues long) that already had experimentally derived Φ values. Their folding rates were also calculated in the process of the Φ value calculation. In addition, only the folding rates were calculated for another set (45 proteins), for which folding rates were experimentally determined, but Φ value analysis was not carried out.

In the next section, our statistical-mechanical model of protein folding is described in detail. The calculated free-energy profiles, folding rates, and Φ values are shown and discussed in Sections 3, 4, and 5, respectively. Because it is difficult to show all the results here, only a part of the results is presented as an illustration. Detailed results are available at http://www.f.waseda.jp/wako /PhiValues/.

2. THE STATISTICAL-MECHANICAL MODEL OF PROTEIN FOLDING

2.1. Assumptions about the Folding Mechanism

The simple statistical-mechanical model of protein folding and unfolding that we used in this study was constructed with the intention of explaining the following mechanism of protein folding (Wako and Saitô 1978a, b; Abe and Gō 1981; Gō and Abe 1981; Abe and Wako 2006, 2009; Wako and Abe 2011). A protein folds in a stepwise manner along the polypeptide chain. In this scheme, at the first stage of folding, short-range interactions are predominant and form small native-like structures such as the α-helix, β-strand, and a turn. Next, these structures grow gradually via medium-range interactions. Finally, these substructures coalesce into the native structure by means of long-range interactions.

The statistical-mechanical model of folding and unfolding of a protein consisting of n residues in the above-mentioned picture is formulated as follows (Abe and Wako 2009; Wako and Abe 2011).

(i) Each amino acid residue is assumed to be in either a native (folded) or random-coil (unfolded) state.
(ii) A protein conformation at any stage in the folding process is represented by a sequence of two types of regions of various sizes, namely a local structure and a random-coil region, arranged alternately along the chain. The term “local structure” has a specific meaning in this model. A local structure and a random-coil region are defined as continuous regions in which all amino acid residues are either in the native state or in the random-coil state, respectively. It should be noted that the number of local structures is not restricted to one in this model, whereas some other models restrict formation of the local structure to one instance.
(iii) The key assumption of this statistical-mechanical model is that only the Gō-type native interactions between amino acid residues within a

local structure are considered (Taketomi et al., 1975). Other interactions, such as those between the residues in different local structures and those within a random-coil region, are ignored.

(iv) As for the free energy within a random-coil region (where no interactions between the residues exist), it is assumed that only the chain entropy—which depends on the number of residues—contributes to the partition function. The random-coil state is the reference state; i.e., its statistical weight is set to unity.

(v) The minimal size of a local structure is four consecutive residues. An amino acid residue i is regarded to be in the native state if a region of four consecutive amino acid residues $(i - 1)$ to $(i + 2)$ adopts the same conformation as the native structure ($i = 2, 3, ..., n - 2$). Otherwise, the amino acid residue is considered to be in the random-coil state. In other words, interactions between amino acid residue i and amino acid residues $i \pm 1$ and $i \pm 2$ are ignored.

2.2. The Partition Function

The partition function for this statistical-mechanical model was obtained by calculating the sum of the statistical weights of the local structures and random-coil regions across all possible arrangements. The partition function can be derived via repetitive use of the following recurrent relation:

$$Z_{1,j} = Z_{1,j-1} + \sum_{m=1}^{j-3} f(m,j)^{-1} \exp\{-\beta E(m,j)\} Z_{1,m+1},$$

$$(j = 4, 5, \dots, n-1, n)$$

$$Z_{1,1} = 0,\ Z_{1,2} = Z_{1,3} = 1, \tag{1}$$

where $\beta = 1/k_B T$, k_B is the Boltzmann constant, and T is the absolute temperature. $Z_{1,j}$ is the auxiliary partition function of a hypothetical protein molecule consisting of amino acid residues 1 to j. By definition, the partition function of the entire protein molecule is $Z_{1,n}$ ($\equiv Z(T)$).

The conformational energy of a local structure consisting of amino acid residues m to j is given as

$$E(m,j) = \sum_{m \le k, l \le j} U(\xi_k, \xi_l)\Gamma_{k,l}\ , \tag{2}$$

where $U(\xi_k, \xi_l)$ is the energy of interaction between amino acid residues ξ_k and ξ_l, the contact matrix $\Gamma_{k,l} = 1.0$ if amino acid residues k and l are in contact in the native conformation; $\Gamma_{k,l} = 0$ otherwise. Because the interactions between neighboring residues and residues separated by one residue are ignored, $\Gamma_{k,l} = 0$ for $|k - l| \le 2$.$U(\xi_k, \xi_l)$ depends on amino acid types ξ_k and ξ_l. $\Gamma_{k,l}$, i.e., the residue pair in contact, depends on the native structure. Consequently, $E(m, j)$ depends on the amino acid sequence and the native structure. In other words, the amino acid sequence and the native structure are taken into account by means of $E(m, j)$ in this model.

The function $f(m, j)$ corresponds to the number of possible conformations of a segment consisting of amino acid residues m to j in the random-coil state; thus, $k_B \ln f(m, j)$ is the chain entropy of the segment in the random-coil state, and $-k_B \ln f(m, j)$ yields the entropy loss of the segment, when the segment forms the local structure.

In our previous studies on lattice proteins, we used values determined by means of the statistical distributions of contacts in real proteins according to Miyazawa and Jernigan (1985) for $U(\xi_k, \xi_l)$ and the formula $f(m, j) = 1.4084 \times (4.750)^{j-m-2}$ obtained for a cubic lattice polymer (Abe and Wako, 2004) for the entropy term. In this chapter, however, we assigned a single value $\varepsilon < 0$ to $U(\xi_k, \xi_l)$, regardless of the amino acid types, ξ_k and ξ_l, in the wild-type protein (homogeneous contact energy approximation) because $E(m,j) = \sum_{m \le k, l \le j} U(\xi_k, \xi_l)\Gamma_{k,l}$, as determined by Miyazawa and Jernigan (1985), does not necessarily satisfy the condition that the native conformation is at the energy minimum in the conformational ensemble. We preliminarily calculated the Φ values at various ε values, from -0.05 to -0.40, and found that the results were nearly independent of ε. Accordingly, we set ε to -0.10 for all calculations in this study. For $f(m, j)$, because we did not have any established functional form of a real protein, we tentatively used the same formula as we did for the lattice protein, i.e., $f(m, j) = 1.4084 \times (4.750)^{j-m-2}$ according to Abe and Wako (2004). No optimization was conducted for the parameters in $f(m, j)$.

An additional consideration with a real-protein problem is how to define a contact between two residues in the native conformation, i.e., the $\Gamma_{k,l}$ value. We defined the contact as follows: if a distance of at least one atom pair in two residues is shorter than a given cutoff distance, D_c, this residue pair is considered in contact. We performed calculations of Φ values with four D_c values: 4.0, 5.0, 5.5, and 6.0 Å. The results were sensitive to the D_c value for some proteins. We will discuss this finding further later, referring to the calculation results.

2.3. Thermodynamic Properties

Because there are no distinctions between conformational energy and enthalpy in this system, we simply regard the conformational energy as enthalpy. For convenience of computation, the energy (enthalpy) of the system, E_h, is expressed as the integer h in units of 0.01, i.e., $E_h = h\varepsilon_0$ and $\varepsilon_0 = 0.01$. Eventually, the partition function Z is derived as a polynomial equation with two variables, t and u, as a function of the temperature T (Abe and Wako 2004, 2009):

$$Z(T) = \sum_{\eta} \sum_{h} \Omega(\eta, h) t^{\eta} u^{h} , \quad (3)$$

where

$$u = \exp(-\beta\varepsilon_0) \quad (4)$$

and t is a dummy parameter introduced to count the number of amino acid residues in the native state, η, and is set to unity in the last result. η varies from 0 to $n - 3$. The coefficient $\Omega(\eta, h)$ for given values of η and h can be calculated using the recurrent equation, Eq. (1).

The partition function $Z(T)$ can be rewritten as follows:

$$Z(T) = \sum_{h} W_1(h, T) u^{h} \quad (5)$$

$$W_1(h, T) = \sum_{\eta=0}^{n-3} \Omega(\eta, h) t^{\eta} \quad (6)$$

or

$$Z(T) = \sum_{\eta=0}^{n-3} W_2(\eta, T) t^{\eta} \quad (7)$$

$$W_2(\eta,T) = \sum_h \Omega(\eta,h) u^h, \tag{8}$$

where $W_1(h, T)u^h$ and $W_2(\eta, T)t^\eta$ are the sums of the statistical weights across all states, with the given enthalpy E_h ($=h\varepsilon_0$) and with the given number of amino acid residues in the native state η, at the temperature T, respectively.

We define the free energy for a given η from Eq. (8) as

$$F(\eta,T) = -k_B T \ln W_2(\eta,T). \tag{9}$$

Eq. (9) is used to calculate the free-energy profile of the proteins studied in the present work.

The partition function, free energy, enthalpy, and entropy at a given temperature T are given by Eqs. (10)–(13) (the dummy parameter t in Eqs. (6) and (7) is set to unity):

$$Z(T) = \sum_{\eta=0}^{n-3} \exp\{-F(\eta,T)/k_B T\} \tag{10}$$

$$F(T) = -k_B T \ln Z(T) \tag{11}$$

$$H(T) = \sum_h E_h W_1(h,T) u^h / Z(T) \tag{12}$$

$$S(T) = \frac{1}{T}\{H(T) - F(T)\} \tag{13}$$

2.4. Folding and Unfolding Rates

The kinetics of the folding and unfolding processes for a protein (such as folding and unfolding rates) were formulated as a motion along a 1D free-energy profile, with the number of amino acid residues in the native state as a reaction coordinate, according to Muñoz and Eaton (1999) and extensive examination by Henry and Eaton (2004). Because we calculated the free-energy profile $F(\eta, T)$ (Eq. (9)), we applied the former method. According to this method—a simple approach that involves solving a system of differential equations describing reversible hopping between adjacent discrete values of reaction coordinates (η and η+ 1 in this study)—the characteristic relaxation rate can be calculated as follows:

$$\frac{1}{k} \propto \tau \equiv \int_0^\infty \frac{\langle \eta \rangle_{\mathrm{eq}} - \langle \eta(t) \rangle}{\langle \eta \rangle_{\mathrm{eq}} - \langle \eta(0) \rangle} dt$$

$$= \{\langle \eta \rangle_{\mathrm{eq}} - \langle \eta \rangle_0\}^{-1} \sum_{j=0}^{n-4} \frac{1}{p_{\mathrm{eq}}(j) s_{j,j+1}} \sum_{i=j+1}^{n-3} \{p_{\mathrm{eq}}(i) - p_0(i)\} \sum_{\eta=0}^{j} p_{\mathrm{eq}}(\eta)(\langle \eta \rangle_{\mathrm{eq}} - \eta) \,. \quad (14)$$

Here, an equilibrium value of η (at temperature T)

$$\langle \eta \rangle_{\mathrm{eq}} = \sum_{\eta=0}^{n-3} \eta \, p_{\mathrm{eq}}(\eta) \quad (15)$$

can be calculated by means of $F(\eta, T)$, where $p_{\mathrm{eq}}(\eta)$ is the probability that a conformation has ηamino acid residues in the native state:

$$p_{\mathrm{eq}}(\eta) = Z^{-1} \exp\{-F(\eta, T)/k_{\mathrm{B}}T\} \,. \quad (16)$$

Here, $\langle \eta \rangle_{\mathrm{eq}}$ and $p_{\mathrm{eq}}(\eta)$ are functions of T, but T is omitted for clarity.

The relaxation rate k is estimated as the mean rate of relaxation of the average number of native residues to its equilibrium value, starting with the entire population in the completely unfolded state, i.e., $\eta = 0$:

$$p_0(\eta) = \delta_{\eta,0} \quad (17)$$

$$\langle \eta \rangle_0 = \sum_{\eta=0}^{N-3} \eta \, p_0(\eta) = 0 \,. \quad (18)$$

The hopping rates from j to $j + 1$ and from $j + 1$ to j are assumed to be

$$s_{j,j+1} = \gamma\left(\frac{p_{\mathrm{eq}}(j+1)}{p_{\mathrm{eq}}(j)}\right)^{\kappa} \text{ and } s_{j+1,j} = \gamma\left(\frac{p_{\mathrm{eq}}(j+1)}{p_{\mathrm{eq}}(j)}\right)^{\kappa-1}, \quad (19)$$

respectively, to satisfy the detailed balances $s_{j,j+1}/s_{j+1,j} = p_{\rm eq}(j+1)/p_{\rm eq}(j)$. According to Henry and Eaton (2004), we set κ to 0.5 and γ to 1 (the choice of κ has little or no effect on the results according to Henry and Eaton (2004)).

The characteristic relaxation rate depends on the initial condition because of approximation. We can consider another initial condition, i.e., all populations are in the native state of $\eta = n-3$:

$$\frac{1}{k} \propto \tau \equiv \int_0^\infty \frac{\langle\eta\rangle_{\rm eq} - \langle\eta(t)\rangle}{\langle\eta\rangle_{\rm eq} - \langle\eta(0)\rangle} dt$$

$$= \{\langle\eta\rangle_{\rm eq} - \langle\eta\rangle_0\}^{-1} \sum_{j=0}^{n-4} \frac{1}{p_{\rm eq}(j+1)s_{j+1,j}} \sum_{i=0}^{j} \{p_{\rm eq}(i) - p_0(i)\} \sum_{\eta=j+1}^{n-3} p_{\rm eq}(\eta)(\langle\eta\rangle_{\rm eq} - \eta) \quad (20)$$

$$p_0(\eta) = \delta_{\eta,n-3} \quad (21)$$

$$\langle\eta\rangle_0 = \sum_{\eta=0}^{n-3} \eta\, p_0(\eta) = n-3 \quad (22)$$

In this estimation, the characteristic relaxation rate for a given temperature is considered to be approximated by the smaller of the two rates. Although it is possible to consider other initial conditions and it is possible that these conditions lead to different relaxation rates, we considered two extreme cases and chose one condition for a given temperature. As a result, $k_{\rm f}$ above $1/T_{\rm m}$ and $k_{\rm u}$ below $1/T_{\rm m}$ are the folding and unfolding rates, respectively. For estimation of $\ln k_{\rm f}$ and $\ln k_{\rm u}$ near the transition temperature $T_{\rm m}$, we simply assumed linearity of the logarithmic rates in relation to the inverse temperature, $\sim 1/T_{\rm m}$. We estimated $\ln k_{\rm f}$ and $\ln k_{\rm u}$ by linearly extrapolating $\ln k_{\rm f}$ from higher to lower $1/T$ and $\ln k_{\rm u}$ from lower to higher $1/T$, respectively. The intersection point of the two extrapolated lines is defined as a transition temperature, $T_{\rm m}$, i.e., $\ln k_{\rm f}(T_{\rm m}) = \ln k_{\rm u}(T_{\rm m})$.

2.5. The Φ Value

We considered a single amino acid substitution for each residue of the protein examined. For a single-amino-acid substitution of the kth residue (for example, amino acid residue ξ_k in the wild-type protein is replaced by amino acid w_k), we simply assumed that $U(\xi_k, \xi_l)$ is transformed to $U(w_k, \xi_l)$ in Eq. (2). Thus, the Φ values of the *k*th residue were calculated for the two cases where $U(w_k, \xi_l) = \varepsilon + 0.01$ and $\varepsilon - 0.01$ (referred to as plus and minus perturbations, respectively), and the results were averaged (recall that $U(\xi_k, \xi_l) = \varepsilon$ for each residue pair in the wild-type protein). The responses to the substitutions were examined on the basis of changes in the logarithmic folding and unfolding rates, $\ln k_f$ and $\ln k_u$, determined in this study. The changes, $\Delta\ln k_f$ and $\Delta\ln k_u$, were used to calculate the Φ value defined by Eq. (23):

$$\Phi = \frac{\Delta\Delta F_{\ddagger\text{-D}}}{\Delta\Delta F_{\text{N-D}}} = \frac{\Delta \ln k_f}{\Delta \ln k_f - \Delta \ln k_u}, \tag{23}$$

where $\Delta\ln k_f = \ln k_f^{mut}(T_m) - \ln k_f^{wild}(T_m)$ and $\Delta\ln k_u = \ln k_u^{mut}(T_m) - \ln k_u^{wild}(T_m)$ because $\ln k_f^{wild}(T_m) = \ln k_u^{wild}(T_m)$ and $\Delta\ln k_f - \Delta\ln k_u = \ln k_f^{mut}(T_m) - \ln k_u^{mut}(T_m)$. The superscripts "wild" and "mut" denote the wild-type protein and mutated protein, respectively.

2.6. The Proteins Examined

The Φ values were calculated for 27 proteins for which Φ value analysis was performed experimentally. In addition, only the free-energy profiles and folding and unfolding rates were calculated for another set (45 proteins), for which folding and unfolding rates were determined experimentally, but Φ value analysis was not performed (the free-energy profiles and folding and unfolding rates for the former 27 proteins were calculated during their Φ value calculations). The Protein DataBank (PDB) entry codes, sizes, and folding types of these proteins are listed in Table 1 along with the references to experimental studies in a footnote. During the selection of the proteins, we used the data from Garbuzynskiy et al., (2004, 2013) as a reference. Our protein set contains some homologous proteins. Because the available data were limited, we used all the data without considering the redundancy in the analyses that follow. PDB entry codes are used as protein names hereafter for convenience.

3. Free-Energy Profile

Protein behavior is controlled by the protein's free-energy landscape, but it cannot be visualized because it spans a multidimensional space. It has to be projected into 1D or 2D space at the expense of details. Figure 1 shows free-energy profiles, $F(\eta, T_m)$, defined by Eq. (9), plotted against the ratio of amino acid residues in the native state, η, at the transition temperature T_m. The free energy at the largest local maximum, i.e., at the transition point, $\eta^{\ddagger}$, is set to zero for every protein in Figure 1. The ratio η is usually regarded as a reaction coordinate of a folding or unfolding process. Figure 1 shows the free-energy profiles of three proteins, 1AYI, 1GXT, and 1TEN, which have a similar size (86, 88, and 90 amino acid residues, respectively) and have different folding types (all α, α+β, or all β, respectively).

The free-energy profile is characterized by two local minima—at η_u and η_f in the denatured and folded states, respectively—and a local maximum at $\eta^{\ddagger}$ in between, i.e., $\eta_u<\eta^{\ddagger}<\eta_f$. The denatured state is an ensemble of non-native structures including secondary-structure elements to be formed fully or partially. Consequently, how denatured a state is depends on the folding type of a protein. Although short-range interactions are dominant in an α-helix, long-range interactions play a central role in a β-sheet. Because the transition state arises from competing conformational energy and chain entropy, its height, $F(\eta^{\ddagger}, T_m) - F(\eta_u, T_m)$, also depends on the folding type. Figure 1 shows that an all-β protein, 1TEN, has deeper local minima, i.e., a higher transition-state barrier, but an all-α protein, 1AYI, has shallower local minima, i.e., a lower transition-state barrier, and an α+β protein, 1GXT, is in an intermediary situation.

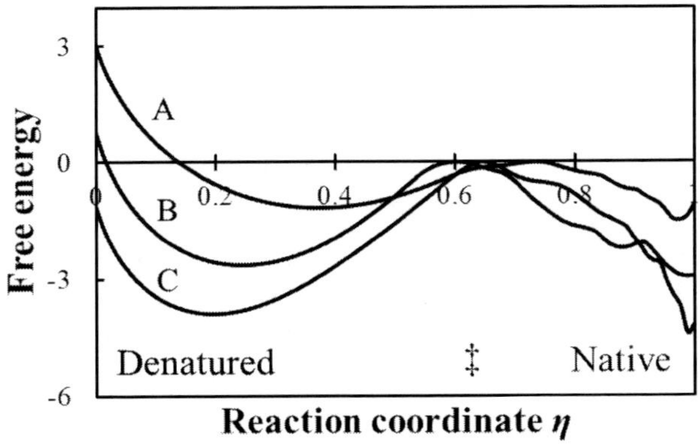

Figure 1. Free-energy profiles of three proteins: (A) 1AYI, (B) 1GXT, and (C) 1TEN.

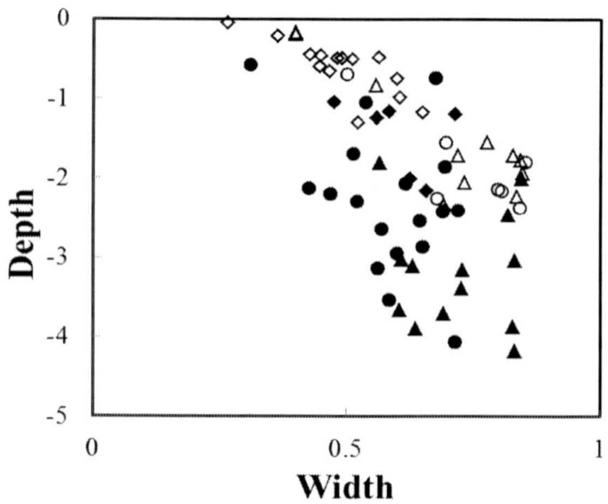

Figure 2. Width and depth of an unfolding region in a free-energy profile of a protein from Table 1. The symbols indicate the folding type: ◇ and ◆, all α; △ and ▲, all β; and ○ and ●, α+β or α/β. The open and closed symbols indicate the protein sizes: smaller than and larger than 80 residues, respectively.

In order to characterize all analyzed proteins in terms of a free-energy profile, we defined "width" and "depth" of an unfolding region. Width is defined as $\eta^{\ddagger} - \eta_{u0}$, where η_{u0} is the intercept of $F(\eta, T_m) = 0$ in the unfolding region (if $F(\eta, T_m) \neq 0$ for any η in the unfolding region, e.g., 1TEN in Figure 1, then $\eta_{u0}$is set to zero). Depth is the difference in free energy: $F(\eta^{\ddagger}, T_m) - F(\eta_u, T_m)$. The width and depth of the unfolding region for the proteins listed in Table 1 are plotted in Figure 2, with classification according to their sizes and folding types.

Obviously, the depth depends on protein size. The smaller the protein (denoted by an open symbol), the smaller the depth is. The folding types also affect the depth: the depth of an all-α protein is smaller than that of all-β proteins, and depths of α+β and α/β proteins are somewhere in between. Dependence of the size and the folding type on the depth are determined by the ratio of long-range interactions to short-range ones in the native structures. This ratio is also related to contact order, i.e., mean separation along a chain between residues making contact in the native structure. The depth is also the height at the transition-state barrier from the denatured state to folded state. Figure 1 is consistent with the well-known fact that the height is strongly related to the entropy loss during formation of a native-like structure, and that formation of long-range interactions is accompanied by a much greater entropy loss in comparison with short-range interactions.

The width of the unfolding region of a free-energy profile depends on a folding type but does not depend strongly on protein size. The smaller width of all-α proteins results from the unfavorable structures corresponding to smaller η, i.e., the complete random-coil structures; the formation of short α-helices is more favorable than formation of the random-coil state near T_m. The width for all-β proteins is relatively greater. This finding indicates that their transition points $\eta^{\ddagger}$ are closer to $\eta = 1$ than those of all-α and α+β proteins. In the competition between a conformational-energy gain and entropy loss, for the all-β proteins, it is necessary to form almost the whole structure at once.

4. The Folding Rate

The folding and unfolding rates are some of the observable properties of a free-energy profile and of a native 3D structure of a protein. A correlation between the folding rate and contact order was shown in many studies (Plaxco et al. 1998, 2000; Gromiha and Selvaraj 2001; Kamagata et al. 2004; Kuznetsov and Rackovsky 2004). It was also shown that an Ising-like statistical-mechanical model, in which interactions are possible only between residues making contact in the native structure, can predict the relative change in rates with similar accuracy (Muñoz and Eaton 1999).

Figure 3 shows the folding rates at a transition temperature, T_m, i.e., $\ln k_f(T_m) = \ln k_u(T_m)$, calculated by means of Eqs. (14) and (20) at $D_c = 5.0$ Å. The good correlation between experimentally observed and theoretically calculated $\ln k_f(T_m)$, which have already been obtained previously by means of a simple statistical-mechanical model by Muñoz and Eaton (1999), are reproduced well with additional data. The correlation coefficient for all proteins is 0.81, and those for the three folding types (all-α proteins, all-β proteins, and α+β and α/β proteins) are 0.67, 0.79, and 0.76, respectively. The correlation coefficients for the proteins larger than 80 residues and smaller than 80 residues are 0.74 and 0.72, respectively. The regression line for all proteins in Figure 3 is $y = 0.64x + 9.95$, where x and y are the theoretically calculated folding rate and experimentally observed folding rate, respectively.

The proteins are classified according to their sizes and folding types in Figures 2 and 3. In general, the smaller the protein, the faster the folding is, and an all-α protein folds faster than an all-β protein does, whereas α+β and α/β proteins are somewhere in between. Because the folding rates were derived on the basis of the free-energy profiles discussed above, the same argumentation is true for the folding rates with respect to the size and folding type as with the free-energy profile.

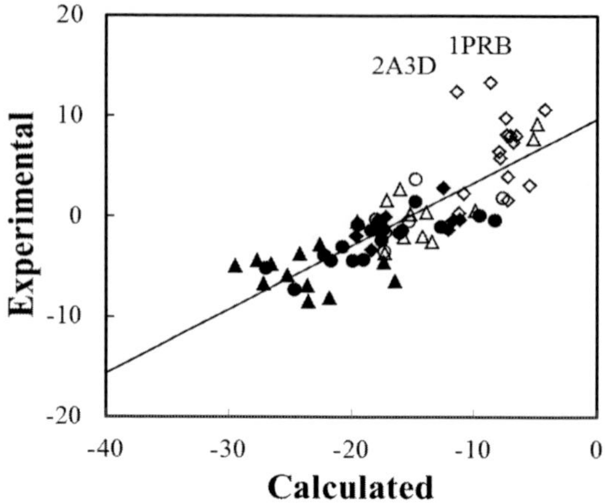

Figure 3. The correlation between experimentally observed and theoretically calculated midtransition folding (unfolding) rates, i.e., ln $k_f(T_m)$ = ln $k_u(T_m)$, at D_c = 5.0 Å. The two proteins with exceptionally high folding rates observed in an experiment, 2A3D and 1PRB, are indicated. See also the caption of Figure 2 for the definitions of the symbols and explanation of the regression line in the main text.

There are two exceptions, proteins 2A3D and 1PRB, as shown in Figure 3. Both are small three-helix bundle proteins. They were reported as ultrafast-folding proteins (Zhu et al. 2003; Wang et al. 2004). It may be necessary to formulate the model for a helix-coil transition more accurately to reproduce the data for such proteins. The correlation coefficients between experimentally observed and theoretically calculated folding rates for all proteins and all-α proteins increase to 0.83 and 0.70, respectively, if we exclude 2A3D and 1PRB from the analysis.

5. ANALYSIS OF Φ VALUES

5.1. Comparison of Calculated and Experimental Φ Values

The Φ values of 27 proteins were calculated by means of Eq. (23). Because Eq. (23) can be rewritten as

$$\Phi = 1/(1 - \Delta \ln k_u / \Delta \ln k_f)\,, \tag{24}$$

$\Delta\ln k_u$ and $\Delta\ln k_f$ should have a different sign for $0 < \Phi < 1.0$. In reality, sometimes $\Phi > 1$ or $\Phi < 0$, i.e., $\Delta\ln k_u$ and $\Delta\ln k_f$ have the same sign. In such cases, some of the calculated Φ values with the perturbations, $U(w_k, \xi_l) = \varepsilon + 0.01$ and $\varepsilon - 0.01$ (see section 2.5), differ substantially. In addition, there is a case where Φ cannot be calculated because a mutated residue has no contact with any other residues in the native structure for a given D_c value (remember that an inter-residue contact is considered only for a pair of residues separated by more than one residue along the chain).

Most of such irregular residues do not have corresponding experimental Φ value data. In other words, a Φ value either could not be or was not obtained experimentally for such residues. The remarkable exceptions are 1TEN and 1FKB. Many of the residues of the two proteins have irregular Φ values when the plus perturbation is imposed. This problem will be discussed below in relation to Figure 9.

The Φ values calculated at D_c = 4.0, 5.0, 5.5, and 6.0 Å were assessed by means of correlation coefficients with the corresponding experimentally observed Φ values; these coefficients are hereafter referred to as γ(Exp, Cal). In this assessment, the residues with an irregular experimental Φ value, i.e., $\Phi < -0.1$ or $\Phi > 1.1$, were excluded from the analysis. If more than one mutation was examined for a given residue in the experiment, a mean Φ value for the residue was used to calculate γ(Exp, Cal). On the other hand, the calculated Φ value was used if it was obtained successfully, even if it was an irregular value. The results are shown in Table 2. For some proteins, the experimental Φ values were obtained under various conditions, but Table 2 shows only one of them. The full data are available at http://www.f.waseda.jp/wako/PhiValues.

In Table 2, the results of Garbuzynskiy et al., (2004) are presented for comparison. In their assessment of their results, those researchers referred only to the experimental data for point mutations that decreased the side-chain volumes and satisfied the condition $0 < \Phi < 1$. When several Φ values satisfying these conditions were obtained for a residue, they were averaged. In the calculations, those researchers considered single-site substitutions with glycine; these substitutions decreased the side-chain volume. The search for optimal parameters was carried out via comparison of calculated and experimental Φ values. Although their statistical-mechanical model for calculation of a Φ value is rather different from our model, their results have a similar trend.

Table 2. Correlation between experimentally observed and theoretically calculated Φ values

Protein	*Nr*[a)]	*Nm*[b)]	Correlation coefficient[c)]					Description[d)]
			4.0	5.0	5.5	6.0	Garbuzynskiy	
1ENH	54	13	0.28	0.43	0.62	0.59	N/A	
1IDY	54	18	0.69	0.62	0.65	0.58	N/A	
1PGB	56	25	0.36	0.31	0.09	−0.02	0.76 (0.74)	
1SRM	56	34	0.63	0.62	0.64	0.65	0.63 (0.63)	
1SHG	57	14	0.23	0.75	0.75	0.74	0.82 (0.81)	pH 7
1FYN	59	9	0.35	0.42	0.34	0.39	N/A	
1SS1	62	31	0.28	0.33	0.71	0.63	N/A	2 M GdmCl
1BF4	63	21	0.31	0.36	0.35	0.32	0.81(0.58)	
3CI2	64	40	0.45	0.37	0.30	0.19	0.46 (0.35)	4 M GdmCl
2PTL	64	46	0.12	0.25	0.28	−0.36[e)]	0.30 (0.26)	Folding
1CSP	67	20	0.49	0.58	0.54	0.50	N/A	H_2O (kinetics)
1UBQ	76	20	−0.03	0.03	0.01	−0.01	N/A	Unfolding
1AYE	78	18	−0.22	−0.22	−0.21	−0.25	N/A	
2ABD	86	16	−0.76	−0.73	−0.75	−0.74	N/A	
1TIU	89	26	0.46	0.45	0.51	0.44	0.66 (0.66)	
1BTB	89	28	−0.09	−0.18	−0.10	−0.18	0.27 (0.20)	
1TEN	90	26	0.61[e)]	0.68[e)]	0.65[e)]	0.66[e)]	0.68 (0.58)	
1TTF	94	20	0.46	−0.21	−0.10	−0.02	−0.22 (−0.12)	1.75 D'
1URN	96	10	0.89	0.94	0.94	0.94	0.90 (−0.71)	β = 0.5
1N88	96	16	0.49	0.59	0.59	0.47	N/A	
1RIS	97	20	0.17	0.10	0.03	0.03	0.49 (0.41)	1–4 M GdmCl
2ACY	98	22	0.37	0.45	0.42	0.42	N/A	
1FKB	107	22	−0.47	−0.40[e)]	−0.50[e)]	−0.47[e)]	0.32 (0.25)	0 M urea
1RNB	109	28	0.39	0.42	0.40	−0.21	0.66 (0.56)	Water
2VIL	126	24	0.46	0.55	0.47	0.45	0.40 (0.35)	
1AZU	126	17	0.65	0.65	0.64	0.59	N/A	0 M GuHCl
3CHY	128	19	0.81	0.15	0.77	0.83	0.67 (0.65)	

a) The number of residues.

b) The number of mutated residues in Φ value analysis. The residues with an irregular Φ value, i.e., $\Phi < -0.1$ or $\Phi > 1.1$, were excluded from the analysis.

c) Correlation coefficients between experimental and calculated Φ values at cutoff distances D_c = 4.0, 5.0, 5.5, and 6.0 Å; the results from Garbuzynskiy et al., (2004) are also shown if available. The values calculated by Garbuzynskiy et al., (2004) are shown for analyses including and excluding hydrogen atoms (the latter value is shown in parentheses).

d) The experimental conditions in an abbreviated form according to the original paper. These data are shown here only if the experiment was carried out under more than one condition. Only one of the conditions is presented here. See the corresponding reference in footnote f) of Table 1 for details of the experiment.

e) The Φ value shown in this table is the mean of two cases, plus and minus perturbations of the energy of an inter-residue interaction, i.e., $U(w_k, \xi_l) = \varepsilon \pm 0.01$ (see section 2.5 for details). It should be noted that this Φ value is the one obtained from the minus perturbation only because the Φ value corresponding to the plus perturbation is quite irregular (see subsection 5.1.6).

Figures 4–9 show the calculated Φ values plotted against the residue number (referred to as a Φ value profile) along with the experimental Φ values for six proteins as examples (the full data including other proteins are also available on the above website). In these figures, the experimental Φ values are sparse (compare N_r and N_m in Table 2) either because the Φ value experiments were performed for a limited number of residues or because the Φ value could not be obtained experimentally even though the experiment was performed for the relevant residue. In contrast, the Φ value can be calculated essentially for any residues except for some irregular residues as described above. Of the six examples, proteins 1SS1 and 3CHY (Figures 4 and 5) have a strong correlation, 1PGD and 1BF4 (Figures 6 and 7) a weak correlation, and 2ABD and 1FKB (Figures 8 and 9) no correlation between experimentally observed and theoretically calculated Φ values. A more detailed discussion of Figures 4–9 is presented in the following subsections.

5.1.1. The Protein 1SS1

Figure 4 shows the results on 1SS1, which is a three-helix bundle protein of 62 residues. This is a case where theoretically calculated Φ values well reproduce the experimentally observed Φ values: $\gamma(\text{Exp, Cal}) = 0.71$ at $D_c = 5.5$ Å (Table 2). The experimental Φ values indicate that the central α-helix, α2, is nearly fully folded; the N-terminal α-helix, α1, is partially formed; and the C-terminal α-helix, α3, is poorly formed. Similarly, the calculated Φ values indicate that α2 is fully folded and α1 and α3 are partially folded. The gradations of the Φ values within α1 and α3 are also reproduced well. Because the formation of α1 and α3 is at a comparable level in the calculated Φ values, the small but significant difference in the formation of α1 and α3 (revealed in the experiments) was not reproduced.

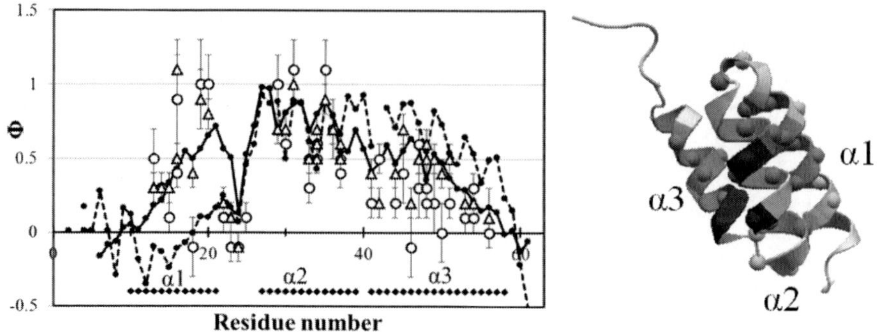

Figure 4. A theoretically calculated Φ value profile, experimentally observed Φ values, and a three-dimensional (3D) structure of the protein 1SS1. (Left) The dashed and solid lines are Φ value profiles at D_c = 4.0 and 5.5 Å, respectively. ○ and △ are the Φ values obtained under different experimental conditions. The strings of a closed symbol at the bottom indicate the location of an α-helix in reference to PDBsum website (de Beer, 2014). In Figures 5–10, open symbols are also used to indicate the location of a β-strand. The residue numbering, −1 to 60, in the Protein DataBank (PDB) data on 1SS1 is shown as 1 to 62. (Right) The protein is presented as a cartoon model. The C_α atom of the residue whose Φ value was obtained in the experiment is shown by a shaded ball: black, gray, and light gray mean Φ > 0.75, 0.75 > Φ > 0.25, and Φ< 0.25, respectively.

In the calculated Φ values, dependence on D_c is observed. As shown in Table 2, some proteins strongly depend on the D_c values, but others do not. Examples of the former case are 1SS1, 2PTL, and 1TTF, and those of the latter case are 1IDY, 1URN, and 1AZU. Because the D_c values reflect inter-residue interactions, these results suggest that proper estimation of inter-residue interactions is necessary to obtain more accurate Φ values. Further research is needed to incorporate D_c into the Φ value calculation. On the basis of such improvement, it would be interesting to see whether the large differences in the calculated Φ values of an α1-helix for D_c = 4.0 and 5.5 Å are related to the variability of the experimental Φ values of this α1-helix.

5.1.2. The Protein 3CHY

Figure 5 shows the results from 3CHY, which is a flavodoxin-like α/β protein of 128 residues. This is the largest of the proteins examined here. 3CHY can be subdivided into two subdomains. The N-terminal one comprises β1-α1-β2-α2-β3, and the C-terminal subdomain β3-α3-β4-α4-β5-α5. The β3-strand belongs to both subdomains. The experimental Φ values suggest that β1-α1-β2 in the N-terminal subdomain is the nucleus around which the rest of the protein folds, whereas the C-terminal subdomain is unstructured. The calculated Φ value profile at D_c = 5.5 Å excellently reproduces the experimental data: γ(Exp, Cal) = 0.77 at D_c = 5.5 Å (Table 2).

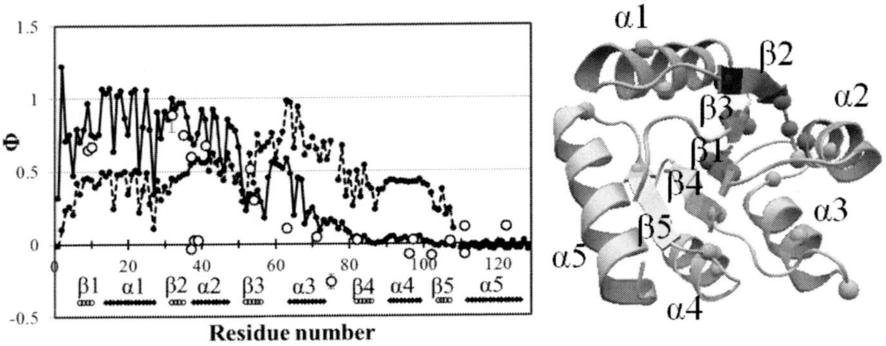

Figure 5. The same analysis as in Figure 4 but for the protein 3CHY. The dashed and solid lines are Φ value profiles at D_c = 5.0 and 5.5 Å, respectively. The residue numbering, 2 to 129, in the PDB data on 3CHY was changed to 1 to128.

In contrast, the Φ value profile at D_c = 5.0 Å is considerably different from that at D_c = 5.5 Å. In fact, as D_c increases from 5.0 to 5.5 Å, the number of inter-residue contacts increases from 414 to 483. Of the 69 additional contacts, 36 and 21 are within the N-terminal and C-terminal halves of 3CHY, respectively, and the remaining 12 are contacts between them. Consequently, the N-terminal subdomain is more strongly stabilized than the C-terminal subdomain as D_c increases from 5.0 to 5.5 Å. In other words, the inter-residue contact information that was obtained at D_c = 5.0 Å is insufficient to reproduce the experimental data.

5.1.3. The Protein 1BF4

Figure 6 shows the results on 1BF4. This is an SH3-fold family protein, as are 1SRM, 1SHG, and 1FYN (the results from these proteins are not shown here). Nevertheless, the correlation coefficient, γ(Exp, Cal), for 1BF4 (0.36 at D_c = 5.0 Å; Table 2) is lower than that for 1SRM, 1SHG, and 1FYN (0.62, 0.75, and 0.42, respectively, at D_c = 5.0 Å; Table 2). The secondary structure sequence of 1BF4 is β1-β2-α1-β3-β4-α2-β5-α3. The difference between experimental and calculated Φ values is significant around the β-hairpin β4-α2-β5. In contrast, the experimental Φ values of the corresponding regions of

1SRM, 1SHG, and 1FYN indicate that they are a part of the folding nucleus, and their calculated Φ values also suggest the same. What is so special about 1BF4? Guerois and Serrano (2000) assumed that three and two glycine residues in the α2-helix and β5-strand, respectively, are responsible for the experimental Φ values of the β-hairpin β4-α2-β5 because these glycine residues imply an additional entropy cost for folding of this β-hairpin. The amino acid sequence features were not taken into consideration in the calculation. Consequently, this result suggests that the protein topology is essential, but amino acid sequence information is also important for accurate calculation of the folding process.

The problem with the β-hairpin β4-α2-β5 is also the problem with the interpretation of an experimentally observed fractional Φ value. This notion will be discussed below from various points of view. Here, we would like to point out a characteristic feature of an Ising-like statistical-mechanical model related to this problem. This is the boundary effect, which is not negligible in a finite system. Because of this effect, the N- and C-terminal regions are less likely to be in the native state than the central region is. In other words, it is more likely that the Φ value is higher in the central region and lower in the N- and C-terminal regions, all else being equal. These phenomena are more pronounced in a smaller system. The boundary effect is not an artifact of the theoretical model but should be taken into account in a real polypeptide. On the other hand, this effect appeared to be somewhat exaggerated in some cases such as 1BF4 although it is difficult to assess it accurately.

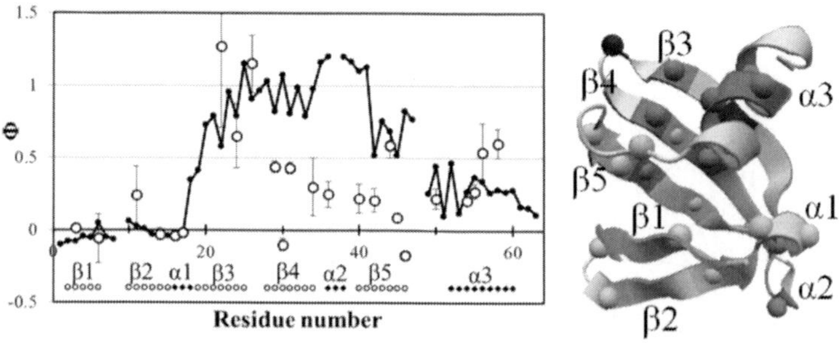

Figure 6. The same analysis as in Figure 4 but for the protein 1BF4. The solid line is a Φ value profile at D_c = 5.0 Å. The residue numbering, 2 to 64, in the PDB data on 1BF4 was changed to 1 to 63.

5.1.4. The Protein 1PGB

1PGB, which is shown in Figure 7, and 2PTL (whose results are not shown here) are IgG-binding domains of protein G and protein L, respectively. Their 3D structures are very similar; both of them comprise two β-hairpins, β1-β2 and β3-β4, and one α-helix, α1, in between. The experimental Φ values, however, suggest that the β-turn of β3-β4 is formed, but that of β1-β2 is disrupted in the transition state of 1PGB, whereas the β-turn of β1-β2 is formed, but that of β3-β4 is disrupted in 2PTL. Obviously, the difference in amino acid sequences (amino acid identity between these proteins is 15%) must be responsible for these phenomena. Because their amino acid sequences were not taken into consideration, the Φ values of the β1-β2 and β3-β4 regions are similar to each other in the calculated Φ value profiles of 1PGB and 2PTL.

Although the calculated Φ values of α1-helices of 1PGB and 2PTL are indicative of full formation in the transition state, the experimental Φ values are fractional. A fractional Φ value generally means that a structure is present but is not fully formed. This topic is controversial. It is interesting to compare these proteins with 3CI2, which has an α-helix located over the β-sheet, as do 1PGM and 2PTL. The experiment with the Φ value shows that the α-helix is within the most ordered element of 3CI2, and the calculated Φ values also indicate this state of affairs. Because the α-helix is mainly stabilized by short-range hydrogen bonding interactions, this helix is likely formed at the early stages of folding. It should be noted that the Φ value is an indicator of interactions between residues not properties of individual residues. Accordingly, it is possible that the Φ values of the residues in an α-helix are small even if the α-helices are formed, as long as the interactions with the residues surrounding each residue under study in the native structure are formed only partially. In addition, it is known that the fractional Φ value is not necessarily proportional to the extent of the interactions formed (Fersht and Sato 2004). Therefore, the difference between the experimental and calculated Φ values for the α1-helices is not easy to understand.

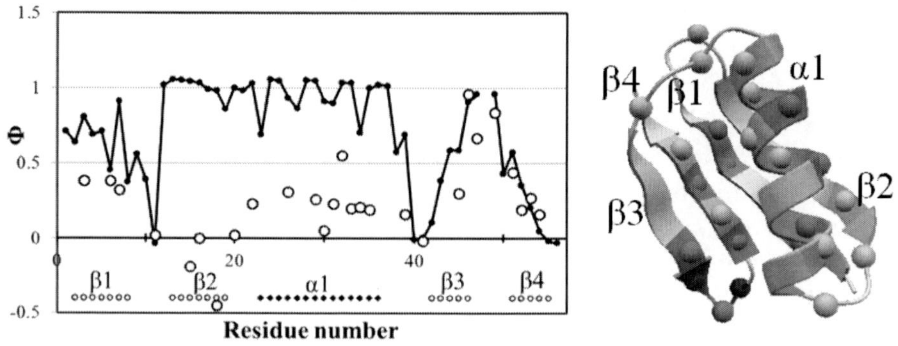

Figure 7. The same analysis as in Figure 4 but for the protein 1PGB. The solid line is a Φ value profile at $D_c = 5.0$ Å.

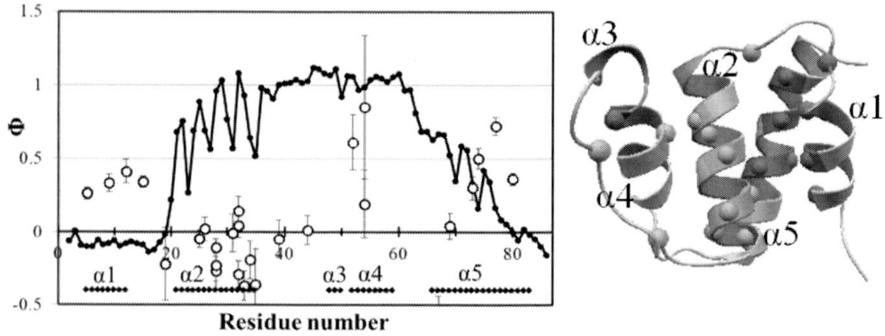

Figure 8. The same analysis as in Figure 4, but for the protein 2ABD. The solid line is a Φ value profile at $D_c = 5.0$ Å.

5.1.5. The Protein 2ABD

Figure 8 shows the results on 2ABD, which is an all-α protein comprising five α-helices: α1 through α5. The experiment shows that the N-terminal and C-terminal α-helices—α1 and α5—form zipper-type specific long-range interactions as a rate-limiting native-like structure, and then the process of folding into the native structure can proceed. In contrast, the calculated Φ value profile indicates that the central three α-helices, α2–α4, form first as a nucleus. Typically, the statistical-mechanical model examined in this chapter cannot deal with such cases: long-range interactions are formed before short- and medium-range interactions. Eventually, the correlation coefficient γ(Exp, Cal) is strongly negative (−0.73 at $D_c = 5.0$ Å; Table 2).

This is a drawback of the statistical-mechanical model presented here. Nonetheless, the results from a well-defined model may provide valuable information on a protein-folding mechanism. In particular, such results elucidate how the framework model and nucleation-condensation mechanism differ and also show that proteins folded via the nucleation-condensation mechanism indeed exist.

5.1.6. The Protein 1FKB

Figure 9 represents a negative result (with the protein 1FKB). The 3D structure of 1FKB is characterized by an antiparallel five-stranded β-sheet (β3 and β4 in Figure 9 may be regarded as one β-strand), which is packed against a small α-helix, α1, and a long loop between β5 and β6 containing a short α-helix, α2. Many long-range interactions stabilize the five-stranded β-sheet. Only one β-hairpin, which is stabilized by medium-range interactions, is formed by β2 and β3. The experimentally observed Φ values and a molecular-dynamics simulation indicated the following: because all the interactions were substantially weaker, a secondary-structure element was not fully formed in the transition state (Fulton et al., 1999). In contrast, the residues with Φ near 1.0 are relatively rare in the calculated Φ value profile at D_c = 4.0 Å, but some of the residues in β3 and α1 may play a role in the formation of nuclei. Incidentally, the correlation coefficient γ(Exp, Cal) is strongly negative (−0.47 at D_c = 4.0 Å; Table 2).

The experimental data seem to be consistent with the nucleation-condensation mechanism of protein folding (Fulton et al., 1999), in which the nucleus does not form a stable structure because the β-sheet requires various long-range interactions to form. In this sense, this is another typical case where our statistical-mechanical model does not work properly.

As mentioned at the beginning of this section, a normal Φ value is between 0 and 1.0. In fact, the calculated Φ values of most residues are in this range. Nevertheless, the calculated Φ values of a considerable number of residues in 1FKB are much greater than 1.0 or much smaller than 0. This phenomenon occurred only in the plus perturbation case (remember that plus and minus perturbations of the inter-residue interaction energy, i.e., $U(w_k, \xi_l) = \varepsilon + 0.01$ and $\varepsilon - 0.01$, respectively, were used in the calculation of Φ values). Accordingly, the correlation coefficients for 1FKB in Table 2 were calculated only from the Φ values with the minus perturbation because they are normal. The same situation was observed for 1TEN whose folding type is β-sandwich of seven strands in two sheets. Nonetheless, the correlation coefficients γ(Exp,

Cal) for 1TEN are remarkably high (Table 2). This is probably because 1TEN contains three β-hairpins (in which a β-sheet is formed with neighboring β-strands), whereas 1FKB contains only one.

The odd results from 1FKB and 1TEN are impressive because both proteins comprise β-structures. Obviously, the long-range interactions could be responsible for these results. The plus perturbation means that long-range interactions between a specific residue and other residues surrounding it are strengthened. The change in these interactions is thought to move the transition state significantly (during competition with the relatively large loss of chain entropy) because of formation of long-range interactions. Because chain entropy was estimated here by the formula used for a lattice protein as a temporary expedient (see section 2.2), further research is needed regarding the method for estimation of chain entropy. It is a challenge to apply our statistical-mechanical model to such complex β-structures.

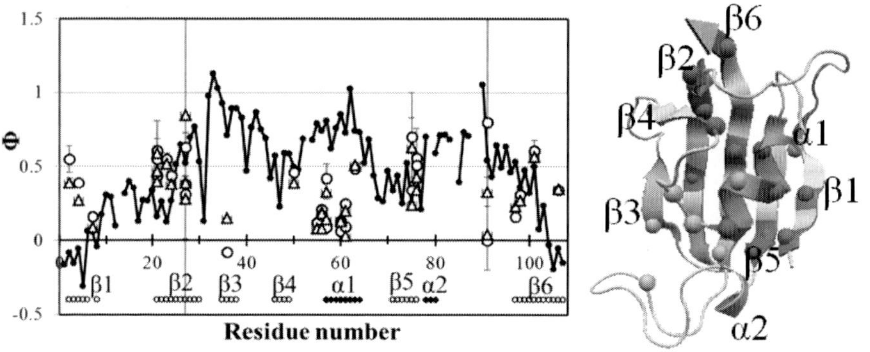

Figure 9. The same analysis as in Figure 4 but for the protein 1FKB. The solid line is a Φ value profile at $D_c = 4.0$ Å.

5.2. Application to Circular Permutation of a Protein

The experiment with Φ value analysis involving a circular permutation of the ribosomal protein S6 (Haglund et al. 2008) may be useful for the discussion of theoretical calculation of a Φ value. A circular permutant is created so that a protein is divided into two fragments, let's say, an N- and C-terminal fragment, by an incision anywhere in the polypeptide, and then the fragments are switched and linked covalently; that is, the N-terminal fragment now follows the C-terminal fragment. The circular permutant that overall has a 3D structure similar to the original structure in spite of its different connectivity (order of secondary-structure elements) is expected to provide useful information on the protein-folding problem.

S6 comprises two α-helices and four β-strands: β1-α1-β2-β3-α2-β4. The PDB code is 1RIS (Lindahl et al., 1994). Five permutants, $P^{13\text{-}14}$, $P^{33\text{-}34}$, $P^{54\text{-}55}$, $P^{68\text{-}69}$, and $P^{81\text{-}82}$, were constructed, where the superscripts indicate the position of the incision. All possible incisions between the secondary-structure elements of S6 were tested. The wild type of S6 is referred to as $S6^{wt}$.

The Φ values for the permutants were calculated by means of the same inter-residue contact information as was used for the native structure of $S6^{wt}$, but we took into consideration the respective connectivity, e.g., 14–97-1–13 for $P^{13\text{-}14}$.

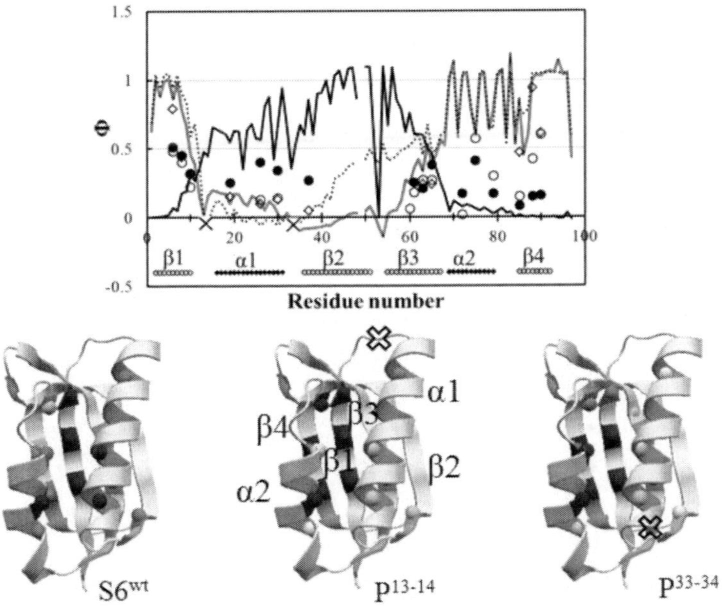

Figure 10. The theoretically calculated Φ value profile, experimentally observed Φ values, and three-dimensional (3D) structures of the wild-type protein 1RIS and its circular permutants. (Top) Black, dotted, and gray lines are Φ value profiles of $S6^{wt}$, $P^{13\text{-}14}$, and $P^{33\text{-}34}$, respectively, at D_c = 5.0 Å. The crosses indicate the incisions, 13-14 and 33-34. The profiles of $P^{54\text{-}55}$, $P^{68\text{-}69}$, and $P^{81\text{-}82}$ are not shown for clarity (see the text in section 5.2). The experimentally observed Φ values are plotted as symbols: ●, ○, and ◇ for $S6^{wt}$, $P^{13\text{-}14}$, and $P^{33\text{-}34}$, respectively. The strings of closed and open symbols at the bottom indicate the locations of an α-helix and β-strand, respectively. (Bottom) The experimental Φ values of $S6^{wt}$, $P^{13\text{-}14}$, and $P^{33\text{-}34}$ are indicated on the 3D structure in a cartoon model (because the 3D structures for $P^{13\text{-}14}$ and $P^{33\text{-}34}$ are not known, the

structure of $S6^{wt}$ is used). The C_α atom of the residue whose Φ value was observed in the experiment is shown as a shaded ball: light gray, gray, and black mean $\Phi < 0.15$, $0.15 < \Phi < 0.3$, and $\Phi > 0.3$, respectively. The crosses indicate positions of the incisions: 13-14 and 33-34 in $P^{13\text{-}14}$ and $P^{33\text{-}34}$, respectively. These experiments and calculations revealed the significance of connectivity of the chain. $S6^{wt}$ and the five permutants are classified according to the β3-strand and α2-nucleus: the secondary-structure elements in β3-strand+α2-nucleus are connected consecutively as β3-α2-β4-β1 in $P^{13\text{-}14}$, $P^{33\text{-}34}$, and $P^{54\text{-}55}$ but separated into the N- and the C-terminal parts in $S6^{wt}$ (β1 and β3-α2-β4, respectively), $P^{68\text{-}69}$ (α2-β4-β1 and β3, respectively), and $P^{81\text{-}82}$ (β4-β1 and β3-α2, respectively). It is possible to classify $S6^{wt}$ and the permutants according to the α1-nucleus and β2-strand; the secondary-structure elements are connected consecutively as β1-α1-β2-β3 in $S6^{wt}$, $P^{68\text{-}69}$, and $P^{81\text{-}82}$ but separated into the N- and C-terminal parts in $P^{13\text{-}14}$ (α1-β2-β3 and β1, respectively), $P^{33\text{-}34}$ (β2-β3 and β1-α1, respectively), and $P^{54\text{-}55}$ (β3 and β1-α1-β2, respectively). The correlation coefficients γ(Exp, Cal) for $S6^{wt}$ and the permutants show that the theoretical calculation can well reproduce the formed β3-α2-β4-β1 and unstructured β1-α1-β2-β3 for $P^{13\text{-}14}$, $P^{33\text{-}34}$, and $P^{54\text{-}55}$, but such calculation cannot reproduce the parallel formation of α1- and α2-nuclei of $S6^{wt}$ and $P^{81\text{-}82}$. The failure to reproduce the early-stage formation of long-range interactions within β3-strand+α2-nucleus separated into N- and C-terminal parts in $S6^{wt}$ and $P^{81\text{-}82}$ is a shortcoming of our statistical-mechanical model as discussed above. On the other hand, the negative γ(Exp, Cal) of $P^{68\text{-}69}$ arises from the difficulty to take into consideration the formation of the α1-nucleus without the formation of the β2-strand in it.

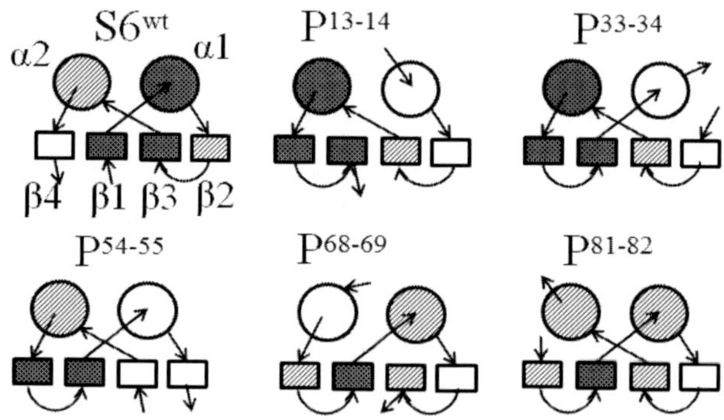

Figure 11. Schematic representation of folding of wild-type and five circular permutants of S6. The α-helix and β-strand are shown as a circle and rectangle, respectively. They are shaded according to average experimentally observed Φ values: white, gray, and black indicate $\Phi < 0.15$, $0.15 < \Phi < 0.3$, and $\Phi > 0.3$, respectively; the figure was made by reference to Haglund et al., (2008).

Figure 10 shows theoretically calculated Φ value profiles of $S6^{wt}$, $P^{13\text{-}14}$, and $P^{33\text{-}34}$ at D_c = 5.0 Å together with the corresponding experimentally observed Φ values. The profiles of $P^{54\text{-}55}$, $P^{68\text{-}69}$, and $P^{81\text{-}82}$ are not shown for clarity; the Φ value profile of $P^{54\text{-}55}$ is very similar to that of $P^{33\text{-}34}$, and the profiles of $P^{68\text{-}69}$ and $P^{81\text{-}82}$ are very similar to that of $S6^{wt}$. The correlation coefficients between experimental and calculated Φ values for $S6^{wt}$, $P^{13\text{-}14}$, $P^{33\text{-}34}$, $P^{54\text{-}55}$, $P^{68\text{-}69}$, and $P^{81\text{-}82}$ are 0.20, 0.89, 0.47, 0.57, −0.16, and −0.20, respectively.

The experiments suggested that the S6 structure is composed of two competing folding nuclei: β1+α1+β3 and β1+α2+β4, referred to as the α1-nucleus and α2-nucleus, respectively (Haglund et al. 2008). The two nuclei fold in parallel in $S6^{wt}$ and $P^{81\text{-}82}$; the α2-nucleus is dominant in $P^{13\text{-}14}$, $P^{33\text{-}34}$, $P^{54\text{-}55}$, and the α1-nucleus is dominant in $P^{68\text{-}69}$. These results are summarized schematically in Figure 11. On the other hand, the calculated Φ values indicate that the α1-nucleus is dominant in $S6^{wt}$, $P^{68\text{-}69}$, and $P^{81\text{-}82}$, whereas the α2-nucleus is dominant in $P^{13\text{-}14}$, $P^{33\text{-}34}$, and $P^{54\text{-}55}$. The calculations, however, suggested that the α1-nucleus should be defined as β1+α1+β2 rather than β1+α1+β3.

Among the circular permutants, a given inter-residue interaction can be short range in one permutant but long range in another. Although some permutants are folded according to a framework model scheme, others are folded according to the nucleation-condensation mechanism. Therefore, the model considered here can well reproduce the experimental Φ values only for some permutants. If more Φ values were available for S6, then we would have a more extensive discussion.

5.3. Problems with Analysis of the Calculated Φ Values

The Φ value calculations are assessed by the correlation coefficients γ(Exp, Cal) between experimental and calculated Φ values, as shown above. γ(Exp, Cal) is, however, not necessary to properly assess our Φ value calculations for the following reasons.

First, the experimental data on Φ values are sparse. Consequently, the large differences between the experimental and calculated Φ values for a few residues can significantly decrease the correlation coefficient. For example, γ(Exp, Cal) = 0.36 of 1BF4 at D_c = 5.0 Å (Table 2) increases to 0.53 only if the experimental Φ value of the 30th residue is excluded from the calculation

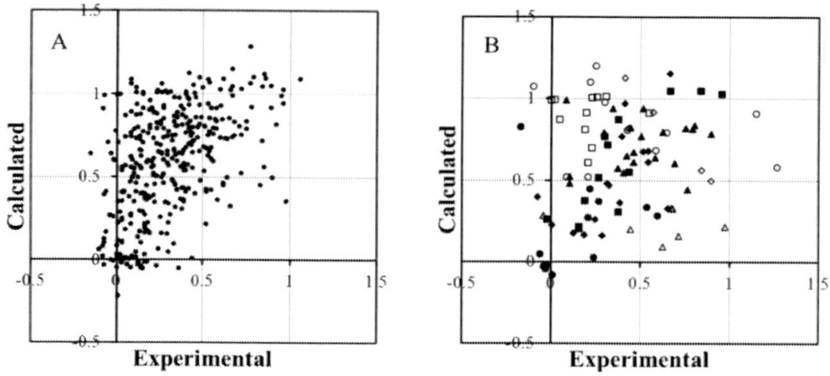

Figure 12. Theoretically calculated Φ values are plotted against experimentally observed ones. (A) The proteins 1IDY, 1PGB, 1SRM, 1SHG, 1FYN, 3CI2, 2PTL, 1CSP, 1TEN, 1TTF, 1URN, 1N88, 2ACY, 1FKB, 1AZU, and 3CHY. (B) ■ and □ mean 1PGB(1–10, 39–56) and 1PGB(11–38), ● and ○ 1BF4(1–20, 47–63) and 1BF4(21–46), ▲ and △ 1BTB(30–89) and 1BTB(1–29), and ◆ and ◇ 1TTF(1–40,55–73) and 1TTF(41–54, 74–94), respectively (D_c = 5.0 Å). See Table 2 for their correlation coefficients.

of γ(Exp, Cal). In addition, it is difficult to analyze the case where a Φ value for a given residue cannot be obtained either experimentally or by calculation although such a case provides some valuable information.

The second problem is a fractional value of Φ. In general, fractional values of Φ are not linear during the formation of an inter-residue interaction, and only the extreme values of zero and 1.0 are fully interpretable per se as being completely denatured-like or completely native-like, respectively (Fersht and Sato 2004). Fractional values can arise from either genuinely weakened interactions in a single transition-state ensemble or from a mixture of states in parallel pathways of folding. Because of these uncertainties in the interpretation of Φ, it was recommended that Φ values be subdivided into three classes: “weak,” “medium,” and “strong” (Fersht 1995). Nevertheless, it is hard to assess the calculated results with such three-grade evaluation.

Third, a good correlation does not always mean that the calculated Φ values are nearly equal to experimental ones. In fact, Figure 12A shows that calculated Φ values are greater than experimental ones even if the correlation coefficients are relatively high. In contrast, Figure 12B shows the cases when the residues in some region have a high correlation and nearly equal values to the experimental ones, but those in another region do not. For such a protein, the correlation coefficient covering all residues is reduced although it is significantly higher for a limited region of the protein. For example, γ(Exp, Cal) = 0.31 for all of 1PGB (Table 2) coexists with 0.75 and 0.35 for some regions: 1PGB(1–10, 39–56) and 1PGB(11–38), respectively; γ(Exp, Cal) = 0.36 for all of 1BF4 (Table 2) coexists with 0.73 and −0.22 for 1BF4(1–20, 47–63) and 1BF4(21–46), respectively.

Finally, it should be noted that the uncertainty in experimental data also makes it difficult to assess the calculated Φ values. As described above, a fractional Φ value and substitution of an amino acid type are major problems in the analysis of Φ values.

CONCLUSION

In this chapter, the Φ values that we obtained by means of a simple statistical-mechanical model of protein folding are discussed and compared with experimentally observed ones. The significance of a study involving a well-defined analytical model is that this approach makes it possible to quantitatively analyze the experimental data instead of conceptually expressed hypotheses about protein folding such as the nucleation-condensation mechanism and framework model. An essential feature of the nucleation-condensation mechanism is concurrent formation of secondary- and tertiary-structure interactions. In contrast, the framework model means hierarchical assembly of structures in which secondary-structure elements are initially formed, independently of tertiary-structure formation; these preformed secondary-structure elements are then thought to coalesce into the native tertiary structure. The model that is presented here is based on the framework model and cannot take into account the concurrent formation of secondary- and tertiary-structure interactions that the nucleation-condensation model is based upon. The above analysis of Φ values shows that the model considered here well reproduces the experimental results for some proteins. In other words, some proteins seem to fold according to the framework model scheme, and others in accordance with the nucleation-condensation mechanism. Thus, there is a need for a novel statistical-mechanical model of protein folding that amalgamates the framework model and nucleation-condensation mechanism in order to solve the protein folding problem.

One limitation of the present model results from the assumption that two residues can interact only if the two residues and the residues in between along the chain are in a native state. This assumption can plausibly introduce into the model the specific interactions that represent the gist of the protein-folding hypotheses. The area accessible to a protein in the conformational space is considered to be restricted in this assumption; in other words, it is assumed that the protein cannot adopt all conformations. The major hypotheses about protein folding, namely, the consistency principle, the minimal-frustration principle, and the funnel hypothesis, are also based on this assumption explicitly or implicitly. The Gō model, in which an inter-residue interaction is allowed only for a residue pair that is in contact in the native structure, is a typical case. We do not believe that real-life protein folding is constrained by this assumption perfectly. In fact, this assumption had to be introduced because otherwise our statistical-mechanical model could not be formulated in an analytically solvable form. If we could extend the model so that the accessible area in the conformational space were wider, e.g., if two residues could interact even if any of the two residues and the residues in between along the chain were not in a native state in the extended model, then formation of the hydrophobic core in remote regions (which is assumed in the nucleation-condensation mechanism) could be considered. A new approach was proposed recently to overcome this problem (Inanami et al. 2014). None of the models of the protein-folding mechanism that have been proposed to date are perfect, but analysis of experimental data with a well-defined model can enrich the understanding of the protein-folding problem. We hope that the statistical-mechanical model of protein folding will be further improved in future studies.

References

Abe, H. and Gō, N. (1981). Noninteracting local-structure model of folding and unfolding transition in globular proteins. II. Application to two-dimensional lattice proteins. *Biopolymers,* 20, 1013–1031.

Abe, H. and Wako, H. (2004). Application of a statistical-mechanical model for protein folding to a three-dimensional lattice protein. *J. Phys. Soc. Jpn,* 73, 1143–1146.

Abe, H. and Wako, H. (2006). Analyses of simulations of three-dimensional lattice proteins in comparison with a simplified statistical mechanical model of protein folding. *Phys. Rev. E*, 74, 011913.

Abe, H. and Wako, H. (2009). Folding/unfolding kinetics of lattice proteins studied using a simple statistical mechanical model for protein folding, I: Dependence on native structures and amino acid sequences. *Physica A,* 383, 3442–3454.

Alm, E. and Baker, D. (1999). Prediction of protein-folding mechanisms from free-energy landscapes derived from native structures. *Proc. Natl. Acad.Sci. USA*, 96, 11305–11310.

Alm, E., Morozov, A.V., Kortemme, T. and Baker, D. (2002). Simple physical models connect theory and experiment in protein folding kinetics. *J. Mol. Biol.,* 322, 463–476.

Alonso, D.O. and Daggett, V. (2000). Staphylococcal protein A: Unfolding pathways, unfolded states, and differences between the B and E domains. *Proc. Natl. Acad. Sci. USA*, 97, 133–138.

Anfinsen, C.B. (1972). The formation and stabilization of protein structure. *Biochem. J.,* 128, 737–749.

Banachewicz, W., Johnson, C.M. and Fersht, A.R. (2011a). Folding of the Pit1 homeodomain near the speed limit. *Proc. Natl. Acad. Sci. USA*, 108, 569–573.

Banachewicz, W., Religa, T.L., Schaeffer, R.D., Daggett, V. and Fersht, A.R. (2011b). Malleability of folding intermediates in the homeodomain superfamily. *Proc. Natl. Acad. Sci. USA,* 108, 5596–5601.

Bruscolini, P. and Naganathan, A.N. (2011). Quantitative prediction of protein folding behaviors from a simple statistical model. *J. Am. Chem. Soc*., 133, 5372–5379.

Bryngelson, J.D., Onuchic, J.N., Socci, N.D. and Wolynes, P.G. (1995). Funnels, pathways and the energy landscape of protein folding: a synthesis. *Proteins,* 21, 167–195.

Bryngelson, J.D. and Wolynes, P.G. (1987). Spin glasses and the statistical mechanics of protein folding. *Proc. Natl. Acad. Sci. USA*, 84, 7524–7528.

Burns, L.L., Dalessio, P.M. and Ropson, I.J. (1998). Folding mechanism of three structurally similar β-sheet proteins. *Proteins,* 33, 107–118.

Calloni, G., Taddei, N., Plaxco, K.W., Ramponi, G., Stefani, M. and Chiti, F. (2003). Comparison of the folding processes of distantly related proteins. Importance of hydrophobic content in folding. *J. Mol. Biol*., 330, 577–591.

Caraglio, M. and Pelizzola, A. (2012). Effects of confinement on thermal stability and folding kinetics in a simple Ising-like model. *Phys. Biol*., 9, 016006.

Chiti, F., Taddei, N., White, P.M., Bucciantini, M., Magherini, F., Stefani, M. and Dobson, C.M. (1999). Mutational analysis of acylphosphatase suggests the importance of topology and contact order in protein folding. *Nat. Struct. Biol.*, 6, 1005–1009.

Choe, S.E., Li, L., Matsudaira, P.T., Wagner, G. and Shakhnovich, E.I. (2000). Differential stabilization of two hydrophobic cores in the transition state of the villin 14T folding reaction. *J. Mol. Biol.*, 304, 99–115.

Choe, S.E., Matsudaira, P.T., Osterhout, J., Wagner, G. and Shakhnovich, E.I. (1998). Folding kinetics of villin 14T, a protein domain with a central β-sheet and two hydrophobic cores. *Biochemistry*, 37, 14508–14518.

Clarke, J., Cota, E., Fowler, S.B. and Hamill, S.J. (1999). Folding studies of immunoglobulin-like β-sandwich proteins suggest that they share a common folding pathway. *Structure,* 7, 1145–1153.

Clarke, J., Hamill, S.J. and Johnson, C.M. (1997). Folding and stability of a fibronectin type III domain of human tenascin. *J. Mol. Biol.*, 270, 771–778.

Cota, E. and Clarke, J. (2000). Folding of β-sandwich proteins: three-state transition of a fibronectin type III module. *Protein Sci.*, 9, 112–120.

Cota, E., Steward, A., Fowler, S.B. and Clarke, J. (2001). The folding nucleus of a fibronectin type III domain is composed of core residues of the immunoglobulin-like fold. *J. Mol. Biol.*, 305, 1185–1194.

Dalessio, P.M. and Ropson, I.J. (2000). β-sheet proteins with nearly identical structures have different folding intermediates. *Biochemistry,* 39, 860–871.

de Beer, T.A., Berka, K., Thornton, J.M. and Laskowski, R.A. (2014). PDBsum additions. *Nucleic Acids Res.* (Database issue), D292–296.

Dill, K.A. and Chan, H.S. (1997). From Levinthal to pathways to funnels. *Nat. Struct. Biol.*, 4, 10–19.

Ferguson, N., Johnson, C.M., Macias, M., Oschkinat, H. and Fersht, A.R. (2001). Ultrafast folding of WW domains without structured aromatic clusters in the denatured state. *Proc. Natl. Acad. Sci. USA*, 98, 13002–13007.

Fersht, A.R. (1995). Characterizing transition states in protein folding: An essential step in the puzzle. *Curr. Opin. Struct. Bio*l., 5, 79–84.

Fersht, A.R. (1997). Nucleation mechanisms in protein folding. *Curr. Opin. Struct. Biol.,* 7, 3–9.

Fersht, A.R., Matouschek, A. and Serrano, L. (1992). The folding of an enzyme. I. Theory of protein engineering analysis of stability and pathway of protein folding. *J. Mol. Biol.*, 224, 771–782.

Fersht, A.R. and Sato, S. (2004). Φ-value analysis and the nature of protein-folding transition states. *Proc. Natl. Acad. Sci. USA*, 101, 7976–7981.

Fowler, S.B. and Clarke, J. (2001). Mapping the folding pathway of an immunoglobulin domain: structural detail from phi value analysis and movement of the transition state. *Structure,* 9, 355–366.

Fulton, K.F., Main, E.R., Daggett, V. and Jackson, S.E. (1999). Mapping the interactions present in the transition state for unfolding/folding of FKBP12. *J. Mol. Biol.,* 291, 445–461.

Galzitskaya, O.V. and Finkelstein, A.V. (1999). A theoretical search for folding/unfolding nuclei in three-dimensional protein structures. *Proc. Natl. Acad. Sci. USA,* 96, 11299–11304.

Garbuzynskiy, S.O., Finkelstein, A.V. and Galzitskaya, O.V. (2004). Outlining folding nuclei in globular proteins. *J. Mol. Biol.*, 336, 509–525.

Garbuzynskiy, S.O., Ivankov, D.N., Bogatyreva, N.S. and Finkelstein, A.V. (2013). Golden triangle for folding rates of globular proteins. *Proc. Natl. Acad. Sci. USA*, 110, 147–150.

Garcia-Mira, M.M., Boehringer, D. and Schmid, F.X. (2004). The folding transition state of the cold shock protein is strongly polarized. *J. Mol. Biol.*, 339, 555–569.

Gianni, S., Guydosh, N.R., Khan, F., Caldas, T.D., Mayor, U., White, G.W., DeMarco, M.L., Daggett, V. and Fersht, A.R. (2003). Unifying features in protein-folding mechanisms. *Proc. Natl. Acad. Sci. USA*, 100, 13286–13291.

Gō, N. (1983). Theoretical Studies of Protein Folding. *Annu. Rev. Biophys. Bioeng.*, 12, 183–210.

Gō, N. and Abe, H. (1981). Noninteracting local-structure model of folding and unfolding transition in globular proteins. I. Formulation. *Biopolymers,* 20, 991–1011.

Gromiha, M.M. and Selvaraj, S. (2001). Comparison between long-range interactions and contact order in determining the folding rate of two-state proteins: application of long-range order to folding rate prediction. *J. Mol. Biol.*, 310, 27–32.

Guerois, R. and Serrano, L. (2000). The SH3-fold family: experimental evidence and prediction of variations in the folding pathways. *J. Mol. Biol.*, 304, 967–982.

Guijarro, J.I., Morton, C.J., Plaxco, K.W., Campbell, I.D. and Dobson, C.M. (1998). Folding kinetics of the SH3 domain of PI3 kinase by real-time NMR combined with optical spectroscopy. *J. Mol. Biol.*, 276, 657–667.

Haglund, E., Lindberg, M.O. and Oliveberg, M. (2008). Changes of protein folding pathways by circular permutation. Overlapping nuclei promote global cooperativity. *J. Biol. Chem.*, 283, 27904–27915.

Hamill, S.J., Steward, A. and Clarke, J. (2000). The folding of an immunoglobulin-like Greek key protein is defined by a common-core nucleus and regions constrained by topology. *J. Mol. Biol.*, 297, 165–178.

Hedberg, L. and Oliveberg, M. (2004). Scattered Hammond plots reveal second level of site-specific information in protein folding: ϕ' ($\beta^{\ddagger}$). *Proc. Natl. Acad. Sci. USA,* 101, 7606–7611.

Henry, E.R. and Eaton, W.A. (2004). Combinatorial modeling of protein folding kinetics: free energy profiles and rates. *Chem. Phys.*, 307, 163–185.

Imparato, A., Pelizzola, A. and Zamparo M. (2007). Ising-like model for protein mechanical unfolding. *Phys. Rev. Lett.*, 98, 148102.

Inanami, T., Terada, T.P. and Sasai, M. (2014). Folding pathway of a multidomain protein depends on its topology of domain connectivity. *Proc. Natl. Acad. Sci. USA*, 111, 15969–15974.

Itoh, K. and Sasai, M. (2004). Dynamical transition and protein quake in photoactive yellow protein. *Proc. Natl. Acad. Sci. USA*, 101, 14736–14741.

Itoh, K. and Sasai, M. (2010). Entropic mechanism of large fluctuation in allosteric transition. *Proc. Natl. Acad. Sci. USA*, 107, 7775–7780.

Itzhaki, L.S., Otzen, D.E. and Fersht, A.R. (1995). The structure of the transition state for folding of chymotrypsin inhibitor 2 analysed by protein engineering methods: Evidence for a nucleation-condensation mechanism for protein folding. *J. Mol. Biol.*, 254, 260–288.

Jemth, P., Day, R., Gianni, S., Khan, F., Allen, M., Daggett, V. and Fersht, A.R. (2005). The structure of the major transition state for folding of an FF domain from experiment and simulation. *J. Mol. Biol.*, 350, 363–378.

Kamagata, K., Arai, M. and Kuwajima, K. (2004). Unification of the folding mechanisms of non-two-state and two-state proteins. *J. Mol. Biol.*, 339, 951–965.

Kim, P.S. and Baldwin, R.L. (1982). Specific intermediates in the folding reactions of small proteins and the mechanism of protein folding. *Annu. Rev. Biochem.*, 51, 459–489.

Kim, D. E., Fisher, C. and Baker, D. (2000). A breakdown of symmetry in the folding transition state of protein L. *J. Mol. Biol.*, 298, 971–984.

Kragelund, B.B., Osmark, P., Neergaard, T.B., Schiødt, J., Kristiansen, K., Knudsen, J. and Poulsen, F.M. (1999). The formation of a native-like structure containing eight conserved hydrophobic residues is rate limiting in two-state protein folding of ACBP. *Nat. Struct. Biol.,* 6, 594–601.

Kubelka, J., Henry, E.R., Cellmer, T., Hofrichter, J. and Eaton, W.A. (2008). Chemical, physical, and theoretical kinetics of an ultrafast folding protein. *Proc. Natl. Acad. Sci. USA*, 105, 18655–18662.

Kuznetsov, I.B. and Rackovsky, S. (2004). Class-specific correlations between protein folding rate, structure-derived, and sequence-derived descriptors. *Proteins*, 54, 333–341.

Lappalainen, I., Hurley, M.G. and Clarke, J. (2008). Plasticity within the obligatory folding nucleus of an immunoglobulin-like domain. *J. Mol. Biol.,* 375, 547–559.

Laurents, D.V., Corrales, S., Elias-Arnanz, M., Sevilla, P., Rico, M. and Padmanabhan, S. (2000). Folding kinetics of phage 434 Cro protein. *Biochemistry,* 39, 13963–13973.

Levinthal, C. (1968). Are there pathways for protein folding? *J. Chim. Phys.,* 65, 44–45.

Lindahl, M., Svensson, L.A., Liljas, A., Sedelnikova, S.E., Eliseikina, I.A., Fomenkova, N.P., Nevskaya, N., Nikonov, S.V., Garber, M.B., Muranova, T.A., Rykonova, A.I. and Amons, R. (1994). Crystal structure of the ribosomal protein S6 from *Thermus thermophilus. EMBO J.,* 13, 1249–1254.

Liu, C., Gaspar, J.A., Wong, H.J. and Meiering, E.M. (2002). Conserved and nonconserved features of the folding pathway of hisactophilin, a β-trefoil protein. *Protein Sci.*, 11, 669–679.

López-Hernández, E. and Serrano, L. (1996). Structure of the transition state for folding of the 129 aa protein CheY resembles that of a smaller protein, CI-2. *Fold. Des.*, 1, 43–55.

Lowe, A.R. and Itzhaki, L.S. (2007). Rational redesign of the folding pathway of a modular protein. *Proc. Natl. Acad. Sci. USA*, 104, 2679–2684.

Manyusa, S. and Whitford, D. (1999). Defining folding and unfolding reactions of apocytochrome b5 using equilibrium and kinetic fluorescence measurements. *Biochemistry,* 38, 9533–9540.

Martinez, J.C. and Serrano, L. (1999). The folding transition state between SH3 domains is conformationally restricted and evolutionarily conserved. *Nat.Struct. Biol.*, 6, 1010–1016.

Matouschek, A., Kellis, J.T. Jr., Serrano, L., Bycroft, M. and Fersht, A.R. (1990). Transient folding intermediates characterized by protein engineering. *Nature*, 346, 440–445.

Matouschek, A., Kellis, J.T., Serrano, L. and Fersht, A.R. (1989). Mapping the transition state and pathway of protein folding by protein engineering. *Nature,* 340, 122–126.

Maxwell, K.L., Wildes, D., Zarrine-Afsar, A., et al., (2005). Protein folding: defining a "standard" set of experimental conditions and a preliminary kinetic data set of two-state proteins. *Protein Sci.*, 14, 602–616.

McCallister, E.L., Alm, E. and Baker, D. (2000). Critical role of β-hairpin formation in protein G folding. *Nat. Struct. Biol.*, 7, 669–673.

Miyazawa, S. and Jernigan, R.L. (1985) Estimation of effective interresidue contact energies from protein crystal structures: quasi-chemical approximation. *Macromolecules,* 18, 534–552.

Muñoz, V. (2014). A simple theoretical model goes a long way in explaining complex behavior in protein folding. *Proc. Natl. Acad. Sci. USA*, 111, 15863–15864.

Muñoz, V. and Eaton, W.A. (1999). A simple model for calculating the kinetics of protein folding from three-dimensional structures. *Proc. Natl. Acad. Sci. USA,* 96, 11311–11316.

Muñoz, V., Lopez, E.M., Jager, M. and Serrano, L. (1994). Kinetic characterization of the chemotactic protein from Escherichia coli, CheY. Kinetic analysis of the inverse hydrophobic effect. *Biochemistry,* 33, 5858–5866.

Murzin, A.G., Brenner, S.E., Hubbard, T. and Chothia, C. (1995). SCOP: a structural classification of proteins database for the investigation of sequences and structures. *J. Mol. Biol.*, 247, 536–540.

Myers, J.K. and Oas, T.G. (2001). Preorganized secondary structure as an important determinant of fast protein folding. *Nat. Struct. Biol.,* 8, 552–558.

Neuweiler, H., Sharpe, T.D., Rutherford, T.J., Johnson, C.M., Allen, M.D., Ferguson, N. and Fersht, A.R. (2009). The folding mechanism of BBL: Plasticity of transition-state structure observed within an ultrafast folding protein family. *J. Mol. Biol.*, 390, 1060–1073.

Nickson, A.A., Stoll, K.E. and Clarke, J. (2008). Folding of a LysM domain: entropy-enthalpy compensation in the transition state of an ideal two-state folder. *J. Mol. Biol.*, 380, 557–569.

Nölting, B. and Agard, D.A. (2008). How general is the nucleation–condensation mechanism? *Proteins,* 73, 754–764.

Nölting, B. and Andert, K. (2000). Mechanism of protein folding. *Proteins,* 41, 288–298.

Northey, J.G., Di Nardo, A.A. and Davidson, A.R. (2002). Hydrophobic core packing in the SH3 domain folding transition state. *Nat. Struct. Biol.*, 9, 126–130.

Otzen, D.E. and Oliveberg, M. (2002). Conformational plasticity in folding of the split β-α-β protein S6: evidence for burst-phase disruption of the native state. *J. Mol. Biol.*, 317, 613–627.

Perl, D., Welker, C., Schindler, T., Schröder, K., Marahiel, M.A., Jaenicke, R. and Schmid, F.X. (1998). Conservation of rapid two-state folding in mesophilic, thermophilic and hyperthermophilic cold shock proteins. *Nat. Struct. Biol.*, 5, 229–235.

Plaxco, K.W., Simons, K.T. and Baker, D. (1998). Contact order, transition state placement and the refolding rates of single domain proteins. *J. Mol. Biol.,* 277, 985–994.

Plaxco, K.W., Simons, K.T., Ruczinski, I. and Baker, D. (2000). Topology, stability, sequence, and length: defining the determinants of two-state protein folding kinetics. *Biochemistry*, 39, 11177–11183.

Portman, J.J., Takada, S. and Wolynes, P.G. (1998). Variational theory for site resolved protein folding free energy surfaces. *Phys. Rev. Lett.*, 81, 5237–5240.

Ptitsyn, O.B. (1987). Protein folding: hypotheses and experiments. *J. Protein Chem.*, 6, 273–293.

Reid, K.L., Rodriguez, H.M., Hillier, B.J. and Gregoret, L.M. (1998). Stability and folding properties of a model beta-sheet protein, *Escherichia coli* CspA. *Protein Sci.*, 7, 470–479.

Riddle, D.S., Grantcharova, V. P., Santiago, J.V., Alm, E., Ruczinski, I. and Baker, D. (1999). Experiment and theory highlight role of native state topology in SH3 folding. *Nat. Struct. Biol.,* 6, 1016–1024.

Roumestand, C., Boyer, M., Guignard, L., Barthe, P. and Royer, C.A. (2001). Characterization of the folding and unfolding reactions of a small β-barrel protein of novel topology, the MTCP1 oncogene product P13. J. *Mol. Biol.*, 312, 247–259.

Sato, S., Religa, T.L., Daggett, V. and Fersht, A.R. (2004). Testing protein-folding simulations by experiment: B domain of protein A. *Proc. Natl. Acad. Sci. USA*, 101, 6952–6956.

Schreiber, G. and Fersht, A.R. (1993). The refolding of cis- and trans-peptidylprolyl isomers of barstar. *Biochemistry,* 32, 11195–11203.

Schymkowitz, J.W., Rousseau, F., Irvine, L.R. and Itzhaki, L.S. (2000). The folding pathway of the cell-cycle regulatory protein p13suc1: clues for the mechanism of domain swapping. *Structure,* 8, 89–100.

Scott, K.A., Batey, S., Hooton, K.A. and Clarke, J. (2004). The folding of spectrin domains I: wild-type domains have the same stability but very different kinetic properties. *J. Mol .Biol.*, 344, 195–205.

Serrano, L., Matouschek, A. and Fersht, A.R. (1992). The folding of an enzyme. III. Structure of the transition state for unfolding of barnase analyzed by a protein engineering procedure. *J. Mol. Biol.*, 224, 805–818.

Settanni, G., Rao, F. and Caflisch, A. (2005). Φ-value analysis by molecular dynamics simulations of reversible folding. *Proc. Natl. Acad. Sci. USA,* 102, 628–633.

Sharpe, T.D., Ferguson, N., Johnson, C.M. and Fersht, A.R. (2008). Conservation of transition state structure in fast folding peripheral subunit-binding domains. *J. Mol. Biol.*, 383, 224–237.

Sivanandan, S. and Naganathan, A.N. (2013). A disorder-induced domino-like destabilization mechanism governs the folding and functional dynamics of the repeat protein IκBα. *PLoS Comput. Biol.*, 9, e1003403.

Spector, S. and Raleigh, D.P. (1999). Submillisecond folding of the peripheral subunit-binding domain. *J. Mol. Biol.*, 293, 763–768.

Taddei, N., Chiti, F., Paoli, P., Fiaschi, T., Bucciantini, M., Stefani, M., Dobson, C.M. and Ramponi, G. (1999). Thermodynamics and kinetics of folding of common-type acylphosphatase: comparison to the highly homologous muscle isoenzyme. *Biochemistry, 38*, 2135–2142.

Taketomi, H., Ueda, Y. and Go, N. (1975). Studies on protein folding, unfolding and fluctuations by computer simulation. I. The effect of specific amino acid sequence represented by specific inter-unit interactions. *Int. J. Pept. Protein Res.*, 7, 445–459.

Tanaka, S. and Scheraga, H.A. (1977). Hypothesis about the mechanism of protein folding. *Macromolecules,* 10, 291–304.

Teilum, K., Thormann, T., Caterer, N.R., Poulsen, H.I., Jensen, P.H., Knudsen, J., Kragelund, B.B. and Poulsen, F.M. (2005). Different secondary structure elements as scaffolds for protein folding transition states of two homologous four-helix bundles. *Proteins*, 59, 80–90.

Ternström, T., Mayor, U., Akke, M. and Oliveberg, M. (1999). From snapshot to movie: ϕanalysis of protein folding transition states taken one step further. *Proc. Natl. Acad. Sci. USA*, 96, 14854–14859.

van Nuland, N.A., Meijberg, W., Warner, J., Forge, V., Scheek, R.M., Robillard, G.T. and Dobson, C.M. (1998). Slow cooperative folding of a small globular protein HPr. *Biochemistry,* 37, 622–637.

Villegas, V., Martínez, J.C., Avilés, F.X. and Serrano, L. (1998). Structure of the transition state in the folding process of human procarboxypeptidase A2 activation domain. *J. Mol. Biol.*, 283, 1027–1036.

Wako, H. and Abe, H. (2011). Folding/unfolding kinetics of lattice proteins by applying a simple statistical mechanical model for protein folding. In:

E.C. Walters (Eds.), Protein Folding (pp. 349–376). New York, NY: Nova Science Publishers.

Wako, H. and Saitô, N. (1978a). Statistical mechanical theory of the protein conformation I. General considerations and the application to homopolymers. *J. Phys. Soc. Jpn*, 44, 1931–1938.

Wako, H. and Saitô, N. (1978b). Statistical mechanical theory of the protein conformation II. Folding pathway for protein. *J. Phys. Soc. Jpn*, 44, 1939–1945.

Wang, M., Tang, Y., Sato, S., Vugmeyster, L., McKnight, C.J. and Raleigh, D.P. (2003). Dynamic NMR line-shape analysis demonstrates that the villin headpiece subdomain folds on the microsecond time scale. *J. Am. Chem. Soc.,* 125, 6032–6033.

Wang, T., Zhu, Y. and Gai, F. (2004). Folding of a three-helix bundle at the folding speed limit. *J. Phys. Chem. B,* 108, 3694–3697.

Wenk, M., Jaenicke, R. and Mayr, E.M. (1998). Kinetic stabilisation of a modular protein by domain interactions. *FEBS Lett.*, 438, 127–130.

Went, H.M. and Jackson, S.E. (2005). Ubiquitin folds through a highly polarized transition state. *Protein Eng. Des. Sel.*, 18, 229–237.

Wilson, C.J. and Wittung-Stafshede, P. (2005). Snapshots of a dynamic folding nucleus in zinc-substituted *Pseudomonas aeruginosa* azurin. *Biochemistry,* 44, 10054–10062.

Wittung-Stafshede, P., Lee, J.C., Winkler, J.R. and Gray, H.B. (1999). Cytochrome b562 folding triggered by electron transfer: approaching the speed limit for formation of a four-helix-bundle protein. *Proc. Natl. Acad. Sci. USA*, 96, 6587–6590.

Zhu, Y., Alonso, D.O., Maki, K., Huang, C.Y., Lahr, S.J., Daggett, V., Roder, H., DeGrado, W.F. and Gai, F. (2003). Ultrafast folding of alpha3D: a de novo designed three-helix bundle protein. *Proc. Natl. Acad. Sci. USA*, 100, 15486–15491.

Zong, C., Wilson, C.J., Shen, T., Wolynes, P.G. and Wittung-Stafshede, P. (2006). Φ-Value analysis of apo-azurin folding: Comparison between experiment and theory. *Biochemistry,* 45, 6458–6466.

In: Advances in Protein Folding Research
Editors: Mathew Hale
ISBN: 978-1-63483-593-0

Chapter 3

PROTEIN FOLDING: TOWARDS A SINGLE MOLECULE THEORY

Yi Fang*
Mathematical Science Institute,
The Australian National University, Canberra, Australia

Abstract

The fundamental physical law governing protein folding is the Thermodynamic Principle: the peptide chain of a protein together with the protein's physiological environment determines the protein's native structure; and the native structure has the minimum Gibbs free energy among all conformations of the protein.

Beginning with a thought experiment indicating that protein folding should be considered in the single molecule level and there must be a general folding force that pushes a protein molecule fold along a folding path, we recall the works leading to the thermodynamics principle.. Arguments to support a single molecule understanding of the thermodynamic principle is given. After preparations including the topology of the conformational space $\mathfrak{X}_{\mathfrak{U}}$ of a molecule $\mathfrak{U}$ and necessary thermodynamic and quantum statistics tools, a Gibbs free energy function $G(\mathbf{X}; \mathfrak{E}_N, \mathfrak{U})$ is derived for a globular protein $\mathfrak{U}$, where $\mathbf{X} \in \mathfrak{X}_{\mathfrak{U}}$, $\mathfrak{E}_N$ is the physiological environment of $\mathfrak{U}$. The derivation relies on making tailor made, nano sized tiny thermodynamic systems $\mathfrak{S}_{\mathbf{X},\mathfrak{E}}$ that contains only one protein

*Email address: yi.fang3@gmail.com.

molecule in the conformation $\mathbf{X}$ and a layer of material particles appear in the environment $\mathfrak{E}$.

Many natural phenomena in protein folding and protein structures can be explained with this Gibbs free energy function and its dependence on environment. Among them are a holistic view of protein structure which explains phenomenon such as the chameleon sequences; why just reducing hydrophobic surface area can produce secondary structures and hydrogen bonds; structure stability; allosteric proteins, docking, enzyme-substrate combination; and denaturation and refolding. The new unified explanations are at least as good as current, *ad hoc* ones, in some cases are much better. Testable predictions are given, first among them is the prediction of protein native structure from the only knowledge of the protein's amino acid sequence. It just becomes a pure mathematical problem of minimisation. Another testable prediction is that changed local environment causes amyloid and prion related diseases.

1. Introduction: A Thought Experiment

The introduction of [41] is entitled as "Molecular biophysics at the twenty-first century: from ensemble measurements to single-molecule detection". In biophysics, single molecule experiments started more than 20 years ago, [46]. Measuring a single molecule is never easy, precise measurement of a single molecule in its natural environment is even impossible at present, as stated in [41], "The ideal biophysical method would be capable of measuring atomic positions in molecules in vivo. It would also permit visualisation of the structures that form throughout the course of conformational changes or chemical reactions, regardless of the time scale involved. At present there is no single experimental technique that yield this information."

On the other hand, mentally developing a single molecule theory based on fundamental physical laws has no obstructions at all, but has never been tried. It is a wonder that why theory lagged behind experiment so much for so long. One of the reasons may be a mental obstruction accumulated from the long history of ensemble measurements and thinking. For example, in biophysics, statistical mechanics is a necessary tool, but whenever an ensemble is considered in biophysics, it is thought as a large quantity of identical material particles. In fact, ensemble as large quantity of identical material particles only fit very few natural phenomena, mentally listing of all possible states of a system is a more general and more useful version of ensemble, see for example, [40]. Especially when we can make Gedanken experiments (thought experiments), to look at what is happening in nature, such as a single protein's folding process, and to contemplate their theoretical explanations.

In a Gedanken experiment, we can follow a newly synthesised protein molecule $\mathfrak{U}$ that is just out of ribosome, with all covalent bonds correctly formed, in an initial conformation that is, we know the precise atomic position of any atom in $\mathfrak{U}$. What is protein folding? The initial conformation will change, i.e., the atomic positions will change, we do not know why and what makes them change, but they are changing. Consider an atom $\mathbf{a}_i$ of the protein, mentally we can record the locus of $\mathbf{a}_i$' positions while it changes with time, it is a curve in $\mathbb{R}^3$, recorded as $\mathbf{x}_i(t)$. Let $\{\mathbf{a}_1, \cdots, \mathbf{a}_n\}$ be all atoms in $\mathfrak{U}$, we will have a curve in $\mathbb{R}^{3n}$, $\mathbf{X}(t) = (\mathbf{x}_1(t), \cdots, \mathbf{x}_n(t))$. We will call it a folding path. It is known that a protein works when it takes a specific conformation called the native structure, in our case, say it is $\mathbf{X}_N = (\mathbf{x}_{N,1}, \cdots, \mathbf{x}_{N,n})$. Continue our Gedanken experiment, the folding path, i.e., the loci of atomic positions $\mathbf{X}(t)$, will stop at $\mathbf{X}_N$, that is, there is a $t_0 > 0$, such that $\mathbf{X}(t_0) = \mathbf{X}_N$.

Now consider any two molecules $\mathfrak{U}_1$, $\mathfrak{U}_2$ of the same protein, with initial conformations $\mathbf{X}_1$ and $\mathbf{X}_2$ respectively, in the same cell of an organism. Although a cell in an organism is crowded with various molecules, ions, etc., there are just a few molecules of the same protein, so we imaging the folding of $\mathfrak{U}_1$ and $\mathfrak{U}_2$ are independently, i.e., their folding paths $\mathbf{X}_1(t; \mathbf{X}_1)$ and $\mathbf{X}_2(t; \mathbf{X}_2)$ never cross to each other. Now, eventually, they will reach the native structure $\mathbf{X}_N$, i.e., there are $t_0^1 > 0$ and $t_0^2 > 0$, such that $\mathbf{X}_1(t_0^1; \mathbf{X}_1) = \mathbf{X}_N$ and $\mathbf{X}_2(t_0^2; \mathbf{X}_2) = C(\mathbf{X}_N)$, where $C : \mathbb{R}^3 \to \mathbb{R}^3$ is a congruence and $C(\mathbf{X}_N) = (C(\mathbf{x}_{N,1}), \cdots, C(\mathbf{x}_{N,n}))$, because the folding paths never cross to each other. This suggests that any conformation $\mathbf{X}$ and any congruence C, $\mathbf{X}$ and $C(\mathbf{X})$ should be treated as the same conformation.

The times t_0^1 and t_0^2 are not necessarily equal, depending how much obstructions met by the two molecules when they fold. It is possible some molecule never reached $\mathbf{X}_N$, but stopped at some other conformation $\mathbf{Y}$, or keep trying endlessly. The known fact is, most molecules finally reached $\mathbf{X}_N$.

The Gedanken experiment hints us, for each protein molecule, there must be a folding path, and the folding paths of different molecule in the same cell are independent to each other. Protein folding is an individual molecule issue. But why they fold to $\mathbf{X}_N$ independently? Are they randomly moving to leave a random trace, or they are pushed by some folding force? As proved by contradiction in [25], randomly moving folding is ruled out. That leaves the folding force. The folding force must be general, because most molecules of the same protein eventually fold to the native structure. The only differences of two different molecules of the same protein are their initial conformations, and the difference in their immediate environments. Thus, the folding force also must be depending on the initial conformation and the immediate environment of each molecule, because they may need different time to reach the native structure, and some may never reach the native structure. For example, there might be some particle very rarely occurring in the cell, but anyway it occurs near by the folding path of a molecule and blocks or delays its folding.

Summarising our Gedanken experiment, we conclude that proteins fold because there is a general folding force pushing each molecule to travel along a folding path to achieve the natives structure. The folding is a single molecule phenomenon because two folding paths simultaneously in the same cell never cross each other. But the folding force must depend on the initial conformation and the immediate environment of a molecule, because different molecule may need different time to fold, and some may never fold to the native structure.

Our task is to figure out what fundamental physical law governs such a folding force and how the folding force acts on a single molecule. That is, we are going to construct a single molecule theory of protein folding.

2. The Thermodynamic Principle of Protein Folding

In trying to establish a single molecule theory of protein folding, it will be beneficial to recall the history of protein folding study, especially the process of from inferring molecular behaviour by ensemble measurements, to the formation of molecular structure concept, the consensus that a protein only works when it is in its native structure, and finally the formulation of the thermodynamic principle of protein folding.

For a long time, proteins were thought as just gigantic agglomeration of small molecules. Protein denaturation was known even earlier than that proteins were recognised as true molecules [43]. In the beginning of the 20th century, protein denaturation was defined as loss of their solubility or coagulation [26]. But coagulated proteins were considered as irreversible, i.e., cannot be reversed to their working state, such as the boiled egg white cannot get back as soluble. Therefore, coagulated proteins were not really interesting. In 1925, it was found that denatured haemoglobin under careful environment management can again be crystallised and even combine with oxygen [2]. With reversibility the protein denaturation became more interesting, what happens when a protein molecule be denatured? Does a chemical reaction happened that changed the molecule to other molecules with different constitute of atomic connection? How to explain the reversibility? If the molecule changed chemically, then it seems that the chemical change can go both ways.

In 1929, after years of experimental research on protein denaturation, Hisen Wu concluded that denaturation does not chemically change the protein molecule [44]. Since [44] is very short (it is just an abstract of a report read before the XIIIth International Congress of Physiology at Boston), we quote the whole abstract here:

"The protein molecule is not to be regarded as a long straight chain but rather as a compact structure. Besides the peptide linkage by which the amino acids are joined 'end to end' there are other kinds of linkages which unite different portions of the chain 'laterally.' These lateral linkages are very labile. The chain may be conceived to fold repeatedly at short intervals forming a three-dimensional network somewhat resembling a crystal lattice in which the atoms are replaced by molecules of amino acids. Denaturation is the breaking up of these labile linkage. Instead of being compact the protein molecule now becomes a 'diffuse' structure. The surface is altered and the interior of the molecule is exposed. This explains the decrease in solubility increase in acid and base binding power, and the change in immunological specify which are known to accompany denaturation.

"Coagulation is the association of protein molecules involving the labile linkage. The ease with which some proteins are coagulated by pure mechanical means, such as shaking and high frequency sound waves, suggests a mechanical explanation of the phenomenon. Besides the possibility of union through the labile linkages on the surfaces, the protein molecules may become entangled as they collide."

We see that by 1929, Hisen Wu had the concept of proteins working in a compact structure, denaturation breaks the structure but not the molecule. The working compact structure is called the native structure today.

Later in [45], Hisen Wu gave more details of his protein denaturation theory, proteins fold to their native structures, physical structure changes explains denaturation. His concept of labile lateral linkages between different portions of the peptide chain, which keep the "3-dimensional network" a delicate stability, was a suggested answer to the question of what made proteins fold to their native structures and stay there. From 1936 till today many think that hydrogen bonds are the labile lateral linkages of Hisen Wu, [27] and [36]. But this view is not the consensus of today because whether in native or other structures, the numbers of hydrogen bonds in the molecule/environment complex (the system $\mathfrak{S}_{\mathbf{X}}$) are the same.

The question remained is: what is the force that pushes proteins to their native structures and stabilise it there?

This question was not answered in principle until 1970's, and the detailed answers are still not known today.

After three decades intensive researches, including the denaturation and refolding experiments conducted by Christian B. Anfinsen and his colleagues on ribonuclease, protein folding and refolding processes are both spontaneous was proved beyond doubts. Therefore, the fundamental physical law governing protein folding must be the second law of thermodynamics. The Thermodynamic Principle, also be called Anfinsen's dogma, but Anfinsen just modestly called it thermodynamic hypothesis as he stated in [1]:

"the three-dimensional structure of a native protein in its normal physiological milieu (solvent, pH, ionic strength, presence of other components such as metal ions or prosthetic groups, temperature, etc.) is the one in which the Gibbs free energy of the whole system is lowest; that is, that the native conformation is determined by the totality of interatomic interactions and hence by the amino acid sequence, in a given environment. In terms of natural selection through the design of macromolecules during evolution, this idea emphasised the fact that a protein molecule only makes stable, structural sense when it exists under conditions similar to those for which it was selected – the so-called physiological state."

There are two understandings of the thermodynamic principle. Let us call them the ensemble understanding and single molecule understanding, differing in the understanding of "the whole system" in the above quotation.

The ensemble understanding is that the "whole system" is an ensemble of identical protein molecules in solvent, when almost all molecules in the ensemble take the natives structure, the whole system has the lowest Gibbs free energy. This understanding is very nature. In fact, protein native structure was inferred from ensemble observations, the Gibbs free energy of the ensemble indeed is the lowest when the ensemble performs biological function. By the structure-function relation, it is inferred that proteins in working must be in native structure. Therefore, when almost all molecules in the ensemble take the natives structure, the whole system has the lowest Gibbs free energy.

The trouble is, environment that is so emphasised in Anfinsen's statement is neglected. The ensemble of identical proteins in working condition and in denatured condition, is actually in two different environments. Gibbs free energy, in the form of $G = E + PV - ST$, depends on environment, difference in environment often can explain difference in Gibbs free energy. Measuring Gibbs free energy of the same ensemble in different environments does not tell us what is the Gibbs free energy differences between the native structure and other conformations of the protein in the same environment. A conformation of a protein molecule folds to the native structure of the protein is a historic process, in nature, this process happens in the same environment, the physiological environment. Thus the ensemble understanding of the thermodynamics principle cannot explain why protein folding happens while the environment does not change.

In summary, what we need is comparing of different conformations in the same environment, not the same ensemble in different environments.

Besides, the ensemble understanding of the thermodynamic principle cannot tell us anything about the geometry of the 3-dimensional native structure, such as the shape and its volume, surface and its area, the interior, etc..

The single molecule understanding is that the "whole system" consisting of all conformations of a protein molecule. In any environment, each conformation has a Gibbs free energy value. The native structure is a conformation that has the lowest Gibbs free energy value among all conformations, in the physiological environment. Therefore, native structure can be characterised as a minimiser of a Gibbs free energy function. To predict proteins native structure from the knowledge of its amino acid sequence is a mathematical task of minimisation. Furthermore, the more we know the Gibbs free energy function, the better we understand the geometry of the 3-dimensional natives structure. There are more advantages of this Gibbs free energy function, for example, the folding force should be the negative of the gradient of this function, the folding path can be identified by the function, etc.

We take the second opinion, not only because the advantages mentioned above, but also because the same environment phenomenon of protein folding in nature leaves no other choice. Besides, perhaps Anfinsen originally really meant to this understanding. His use of "the native conformation is determined by the totality of interatomic interactions and hence by the amino acid sequence, in a given environment." hints that the main folding force is the 'the totality of interatomic interaction", therefore, his wording of intra-molecule, instead of inter-molecule, suggests that we should take a single molecule understanding of the thermodynamic principle.

We first discuss the topology of the conformation space $\mathfrak{X}_{\mathfrak{U}}$ of a molecule $\mathfrak{U}$, its boundary $\partial\mathfrak{X}_{\mathfrak{U}}$, its variable exchange, etc. Then we briefly state the thermodynamic and quantum statistical mechanics necessary for our derivation of the Gibbs free energy function $G(\mathbf{X}; \mathfrak{E}_N, \mathfrak{U})$, where $\mathfrak{E}_N$ is the physiological environment of $\mathfrak{U}$. By proofing that $G(\mathbf{X}; \mathfrak{E}_N, \mathfrak{U})$ is lower semi-continuous and the range of it on $\mathfrak{X}_{\mathfrak{U}}$ is the same as the range of it on a compact set, the existence of minimisers are proved. Furthermore, we can prove that $G(\mathbf{X}; \mathfrak{E}_N, \mathfrak{U})$ can be constructed as a smooth function, therefore, we can use the gradient of it to do minimisation and demonstrate the folding force and folding path.

3. The Conformation Space $\mathfrak{X}_{\mathfrak{U}}$ of a Molecule $\mathfrak{U}$

Let $\mathfrak{U}$ be a molecule with n atoms $\{\mathbf{a}_1, \cdots, \mathbf{a}_n\}$. The molecule $\mathfrak{U}$ occupies a space in $\mathbb{R}^3$, so every atom's centre, the atomic centre, is a point in $\mathbb{R}^3$. A conformation is a list the n atomic centers $\mathbf{x}_i \in \mathbb{R}^3$, such that $\mathbf{X} = (\mathbf{x}_1, \cdots, \mathbf{x}_n) \in \mathbb{R}^{3n}$. Let $\mathfrak{X}_{\mathfrak{U}} \subset \mathbb{R}^{3n}$ be the set of all conformations of $\mathfrak{U}$. But because we have to respect the bond angles and bond lengths of $\mathfrak{U}$, the conformation set $\mathfrak{X}_{\mathfrak{U}}$ is not the whole $\mathbb{R}^{3n}$.

3.1 Bond Angles

Let $\bullet$ be the inner product in $\mathbb{R}^3$ and $\mathbf{X} = (\mathbf{x}_1, \cdots, \mathbf{x}_n)$ be a conformation. If $\mathbf{x}_i$ has a bond with both $\mathbf{x}_j$ and $\mathbf{x}_k$, then the bond angle $\alpha_{ij,ik}$ is defined as (see FIGURE 1)

$$0 \le \alpha_{ij,ik}(\mathbf{x}_i, \mathbf{x}_j, \mathbf{x}_k) := \arccos\left(\frac{(\mathbf{x}_j - \mathbf{x}_i) \bullet (\mathbf{x}_k - \mathbf{x}_i)}{|\mathbf{x}_j - \mathbf{x}_i||\mathbf{x}_k - \mathbf{x}_i|}\right) \le \pi.$$

For amino acid molecules, as well as the peptide bond in proteins, there are standard bond angles $\alpha_{ij,ik}$ which should be respected in any conformation $\mathbf{X}$. Therefore, if $\mathfrak{U}$ is a protein molecule, there are standard bond angles $\gamma_{ij,ik}$ such that any conformation $\mathbf{X}$ of $\mathfrak{U}$ should satisfy

$$\gamma_{ij,ik} - \beta_{ij,ik} \le \alpha_{ij,ik} \le \gamma_{ij,ik} + \beta_{ij,ik}. \tag{1}$$

where $\beta_{ij,ik}$'s are positive constants.

Table 1. Average bond lengths in proteins. 1Å= 10^{-10}m

Peptide	Average length	Single	Average length	Double	Average length
C_α - C	1.51 Å	C - C	1.54 Å		
C - N	1.33 Å	C - N	1.48 Å		
N - C_α	1.46 Å	C - O	1.43 Å	C = O	1.24 Å

3.2 Bond Lengths

The distance between two points $\mathbf{x} = (x_1, x_2, x_3) \in \mathbb{R}^3$ and $\mathbf{y} = (y_1, y_2, y_3) \in \mathbb{R}^3$ is $|\mathbf{x} - \mathbf{y}| = \sqrt{\sum_{i=1}^3 (x_i - y_i)^2}$.

Let $\mathbf{X} = (\mathbf{x}_1, \cdots, \mathbf{x}_n)$ be a conformation. For any bond formed by $\mathbf{a}_i$ and $\mathbf{a}_j$, the bond length in $\mathbf{X}$ is $r_{ij} := |\mathbf{x}_i - \mathbf{x}_j| > 0$.

There are standard bond lengths b_{ij} for various bonds. TABLE 1 shows some of the average bond lengths appear in proteins. The conformation $\mathbf{X}$ has to respect these standard bond lengths, that is, if $\mathbf{a}_i$ and $\mathbf{a}_j$ form a covalent bond,

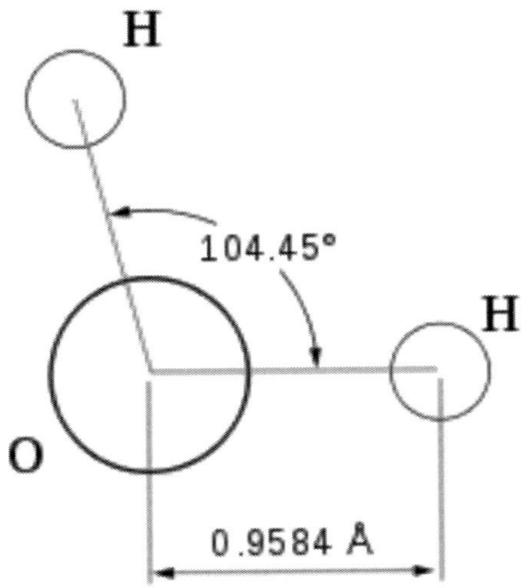

Figure 1. Bond angle and bond length.

then

$$b_{ij} - \delta_{ij} \leq |\mathbf{x}_i - \mathbf{x}_j| \leq b_{ij} + \delta_{ij}, \tag{2}$$

where δ_{ij}'s are positive constants.

3.3 The Topology of $\mathfrak{X}_{\mathfrak{U}}$

The distance of $\mathbf{X} = (\mathbf{x}_1, \cdots, \mathbf{x}_n)$ and $\mathbf{Y} = (\mathbf{y}_1, \cdots, \mathbf{y}_n)$ in $\mathbb{R}^{3n}$ is

$$|\mathbf{X} - \mathbf{Y}| = \sqrt{\sum_{i=1}^{n} |\mathbf{x}_i - \mathbf{y}_i|^2}, \quad \mathbf{x}_i, \mathbf{y}_i \in \mathbb{R}^3, \quad , i = 1, \cdots, n.$$

The open (3n-dimensional) ball in $\mathbb{R}^{3n}$, centred at $\mathbf{X}$ with radius $r > 0$ is

$$B_r(\mathbf{X}) = \{\mathbf{Y} \in \mathbb{R}^{3n} : |\mathbf{Y} - \mathbf{X}| < r\} \subset \mathbb{R}^{3n}.$$

A set $U \subset \mathbb{R}^{3n}$ is open if it is a union of open balls. A set $C \subset \mathbb{R}^{3n}$ is closed if $\mathbb{R}^{3n} \backslash C$ is open, where

$$\mathbb{R}^{3n} \backslash C = \{\mathbf{X} \in \mathbb{R}^{3n} : \mathbf{X} \notin C\}.$$

This is the Euclidean topology of $\mathbb{R}^{3n}$. The $\mathfrak{X}_{\mathfrak{U}} \subset \mathbb{R}^{3n}$ has the inherit topology from $\mathbb{R}^{3n}$, i.e., $O \subset \mathfrak{X}_{\mathfrak{U}}$ is open if and only if there is an open $U \subset \mathbb{R}^{3n}$ such that $O = U \cap \mathfrak{X}_{\mathfrak{U}}$.

3.4 $\mathfrak{X}_{\mathfrak{U}} \subset \mathbb{R}^{3n}$ is a Closed Subset

Let $\mathbf{X} = (\mathbf{x}_1, \cdots, \mathbf{x}_n) \in \mathbb{R}^{3n}$, the functions:

$$f_{ij}(\mathbf{X}) = |\mathbf{x}_i - \mathbf{x}_j| : \mathbb{R}^{3n} \to \mathbb{R}, \quad g_{ij,ik}(\mathbf{X}) = \alpha_{ij,jk}(\mathbf{x}_i, \mathbf{x}_j, \mathbf{x}_k) : \mathbb{R}^{3n} \to \mathbb{R},$$

for all possible $1 \leq i, j, k \leq n$, are continuous. Thus the following sets are closed,

$$\{\mathbf{X} \in \mathbb{R}^{3n} : b_{ij} - \delta_{ij} \leq f_{ij}(\mathbf{X}) \leq b_{ij} + \delta_{ij}\},$$

$$\{\mathbf{X} \in \mathbb{R}^{3n} : \gamma_{ij,ik} - \beta_{ij,ik} \leq g_{ij,ik}(\mathbf{X}) \leq \gamma_{ij,ik} + \beta_{ij,ik}\},$$

for all i, j such that $\mathbf{a}_i$ and $\mathbf{a}_j$ form a covalent bond, and all i, j, k such that $\mathbf{a}_i$ forms covalent bonds with both $\mathbf{a}_j$ and $\mathbf{a}_k$. Hence $\mathfrak{X}_{\mathfrak{U}}$ is the intersection of closed sets, therefore, closed.

The boundary of $\mathfrak{X}_{\mathfrak{U}}$, $\partial\mathfrak{X}_{\mathfrak{U}}$, is the union of the following sets:

$$f_{ij}^{-1}(\{b_{ij}+\delta_{ij}\}), \quad f_{ij}^{-1}(\{b_{ij}-\delta_{ij}\}), \quad \mathbf{a}_i \text{ bonds } \mathbf{a}_j; \tag{3}$$

and

$$g_{ij,ik}^{-1}(\{\gamma_{ij,ik}+\beta_{ij,ik}\}), \quad g_{ij,ik}^{-1}(\{\gamma_{ij,ik}-\beta_{ij,ik}\}), \tag{4}$$

such that $\mathbf{a}_i$ forms covalent bonds with both $\mathbf{a}_j$ and $\mathbf{a}_k$.

Define the interior of $\mathfrak{X}_{\mathfrak{U}}$, $\mathfrak{X}_{\mathfrak{U}}^{\circ} = \mathfrak{X}_{\mathfrak{U}} \backslash \partial\mathfrak{X}_{\mathfrak{U}}$. The $\mathfrak{X}_{\mathfrak{U}}^{\circ}$ is an open set in $\mathbb{R}^{3n}$, so is open in $\mathfrak{X}_{\mathfrak{U}}$. Indeed, by definition of interior, if $\mathbf{X} \in \mathfrak{X}_{\mathfrak{U}}^{\circ}$, then there is an open ball $B_{r_{\mathbf{X}}}(\mathbf{X})$ such that $B_{r_{\mathbf{X}}}(\mathbf{X}) \subset \mathfrak{X}_{\mathfrak{U}}^{\circ}$, thus

$$\mathfrak{X}_{\mathfrak{U}}^{\circ} = \cup_{\mathbf{X}\in\mathfrak{X}_{\mathfrak{U}}^{\circ}} B_{r_{\mathbf{X}}}(\mathbf{X}),$$

$\mathfrak{X}_{\mathfrak{U}}^{\circ}$ is an open set in $\mathbb{R}^{3n}$, therefore, an open set in $\mathfrak{X}_{\mathfrak{U}}$.

Let $U \subset \mathbb{R}^{3n}$ be an open set, then we can talk smoothness of a function $f : U \to \mathbb{R}$, i.e., f has continuous first partial derivatives. Thus we can always discuss the smoothness of a function $f : \mathfrak{X}_{\mathfrak{U}}^{\circ} \to \mathbb{R}$. And if f can be extended to a open set $U \supset \mathfrak{X}_{\mathfrak{U}}$, we can also talk smoothness of f on $\mathfrak{X}_{\mathfrak{U}}$.

3.5 Atomic Space

Given a conformation $\mathbf{X}$, it is obvious that each atom $\mathbf{a}_i$ in $\mathfrak{U}$ occupies a space $B(\mathbf{x}_i) \subset \mathbb{R}^3$. Classical description of $B(\mathbf{x}_i)$ is the van der Waals ball

$$B(\mathbf{x}_i) = B(\mathbf{x}_i, r_i) = \{\mathbf{x} \in \mathbb{R}^3 : \ |\mathbf{x} - \mathbf{x}_i| \leq r_i\},$$

where $r_i > 0$ is the van der Waals radius of the atom $\mathbf{a}_i$. More accurate description of $B(\mathbf{x}_i)$ is obtained by the conception of **atomic surface** defined by the electronic density function ρ of the quantum molecular theory of atoms in molecules, see [3]. Whichever definition of $B(\mathbf{x}_i)$, we define

$$P_{\mathbf{X}} = \cup_{i=1}^{n} B(\mathbf{x}_i) \subset \mathbb{R}^3, \tag{5}$$

to present a 3-dimensional shape of the conformation $\mathbf{X} \in \mathbb{R}^{3n}$.

Remark 3.1. *Classical depiction of the shape of an atom is a ball, or hard sphere as called in chemistry, with the van der Waals radius. We could use the overlapping balls $P_{\mathbf{X}} = \cup_{i=1}^{n} B(\mathbf{x}_i, r_i)$ to present 3-dimensional shape of the conformation $\mathbf{X}$. It is called a CPK or space-filling model. We used more* *general presentation in (5) to be more flexible in the derivation of the Gibbs free energy function.*

Recent experiments and theoretic arguments claim that the hard sphere model of atoms is very faithful to the atomic shape, or an atom behaves physically and chemically more like a hard sphere than a wave function, see [37], [38], and [28]. But by the theory of atoms in molecules of Bader [3], the shape of individual atom in a molecule occupies a region bounded by atomic surfaces that are defined by the gradient of electronic density function. These atomic surfaces depend on the domain in which the Schrödinger equation is solved, when the domain is unbounded the atomic surfaces are also unbound. Hence, treating an atom as a hard sphere is not a good approximation perhaps. However, our real concern is the 2-dimensional compact boundary surface of a conformation's 3-dimensional shape, not each individual atom's shape. The truth is that the cutoff of electron density $\rho \geq 0.001au$ *([3] and [31]), which reflects the space really occupied by the molecule, is very like the boundary shape of a bunch of overlapping balls,* $\partial P_{\mathbf{X}}$*, i.e., the van der Waals surface. Therefore, mathematically we can take atoms as closed balls and will still very faithfully represent the 3-dimensional shape of molecular structures. Later we will take CPK model to guarantee that the derived Gibbs free energy function* $G(\mathbf{X}; \mathfrak{E}_N, \mathfrak{U})$ *smooth.*

3.6 Hydrophobicity Classes

Proteins are huge, emphatic polymers, i.e., the electronic distribution of them is highly uneven. In fact, even within an amino acid, the electronic distribution is already uneven and these uneven charges are carried out in the protein molecules. Because proteins are huge, a protein molecule has many moieties (or atomic groups) with semi-independent physic-chemical properties inherited from amino acids. For example, the main chain amino group NH_2 keeps its ability to form hydrogen bonds. An important property of these moieties in a protein molecule $\mathfrak{U}$ is that they have different levels of hydrophobicity (affinity to water, for example, forming hydrogen bonds) and the space distribution of these moieties in $P_{\mathbf{X}_N}$ of the native structure $\mathbf{X}_N \in \mathfrak{X}_{\mathfrak{U}}$ helps, even determines, the protein's native structure. The thermodynamic principle says that amino acid sequence and environment of a protein determine the protein's native structure. And by current consensus the biological function of a protein relies on the protein's native structure. Combine these together, we infer that the amino acid sequence of the protein and its environment determine the space distribution of these moieties in the natives structure. Therefore, we should pay attention to these moieties and distinguish them by their hydrophobic levels, strongly hydrophobic, weakly hydrophobic, weakly hydrophilic, strongly hydrophilic, etc.. Suppose there are total $k > 1$ classes of hydrophobic levels among all moieties of a protein. Because an atom of the protein belongs only one moiety, each atom in a protein can be assigned a hydrophobicity level, the level of the moiety that contains the atom. For example, we may assume that the classification is as in [12] where there are $k = 5$ classes, C, O/N, O^-, N^+, S. Unlike in [12], we also classify every hydrogen atom into one of the k hydrophobicity classes. Thus we will divide atoms of a protein molecule $\mathfrak{U}$ into k hydrophobic classes H_i, $1 \leq i \leq k$, such that,

$$\{\mathbf{a}_1, \mathbf{a}_2, \cdots, \mathbf{a}_n\} = \cup_{i=1}^{k} H_i, \quad H_i \cap H_j = \emptyset, \quad i \neq j.$$

The hydrophobicity level division not only will help derive a Gibbs free energy function, but also help to distinguish different proteins. To distinguish means that if $\mathfrak{U}_1$ and $\mathfrak{U}_2$ are two different proteins with the same number of atoms, say n. Then the two topological space $\mathfrak{X}_{\mathfrak{U}_1}$ and $\mathfrak{X}_{\mathfrak{U}_2}$ are different.

To derivate the Gibbs free energy function $G : \mathfrak{X}_{\mathfrak{U}} \to \mathbb{R}$, beginning from the peptide chain of $\mathfrak{U}$, $P = A_1 \cdots A_m$, supposing that $\mathfrak{U}$ has m residues, we have to construct a series objectives, all only depend on the peptide chain P, for example, $\mathfrak{X}_{\mathfrak{U}}$. We want these constructions can distinguish proteins, i.e., for different $\mathfrak{U}_1$, $\mathfrak{U}_2$, $\mathfrak{X}_{\mathfrak{U}_1} \neq \mathfrak{X}_{\mathfrak{U}_2}$. And want to extend this distinguishing to all peptide sequences P, either or not P is the peptide sequence of a protein.

The hardest distinguishing task is that given a peptide chain $P = A_1 \cdots A_m$, let m_i, $i = 1, 2, \cdots, 20$, be the number of times of amino acid i appears in P, shuffling the amino acid A_i's around we will have $S(P) = \frac{m!}{\Pi_{j=1}^{20} m_j!}$ different peptide chains, each one has a conformation space. The worst scenario is that the conformation space cannot really distinguish them, then conformation space plus the hydrophobicity classes will, because that hydrophobicity classes distribution in a conformation will be altered if the peptide chain is reshuffled. For example, two atoms, say both are hydrogen, will not be identical, they may belong to different hydrophobicity classes. Thus all peptide chains will be distinguished by atomic number n, peptide chain length m, the conformation space, and the space distributions of the hydrophobicity classes. Since all of these are counted in our desired Gibbs free energy function, $G(\mathbf{X}; \mathfrak{C}_N, \mathfrak{U})$, two different peptide chains will have different Gibbs free energy functions.

Remark 3.2. *The definition of the hydrophobicity classes is independent of conformations. All macro-molecules have their own hydrophobicity classes.*

3.7 Equi-DA Classes

We define a equivalence class in $\mathfrak{X}_{\mathfrak{U}}$. Two conformations in $\mathfrak{X}_{\mathfrak{U}}$ in the same Equi-DA class if and only if they have the same bond angles and lengths. Denote $[\mathbf{X}]$ as the Equi-AD class containing $\mathbf{X}$. Note that if $\mathbf{X} \in \mathfrak{X}_{\mathfrak{U}}^{\circ}$, then $[\mathbf{X}] \subset \mathfrak{X}_{\mathfrak{U}}^{\circ}$, and if $\mathbf{X} \in \partial\mathfrak{X}_{\mathfrak{U}}$, then $[\mathbf{X}] \subset \partial\mathfrak{X}_{\mathfrak{U}}$.

Since $[\mathbf{X}] \subset \mathbb{R}^{3n}$ is the intersection of sets like in (3) and (4) (only with $f_{ij}^{-1}(\{x\})$, $b_{ij} - \delta_{ij} \leq x \leq b_{ij} + \delta_{ij}$, etc.), $[\mathbf{X}]$ is a closed set in $\mathbb{R}^{3n}$, hence is also closed in $\mathfrak{X}_{\mathfrak{U}}$.

3.8 Fast Changing Conformations within the Same Equi-DA Class

$\mathbf{X} \in \mathfrak{X}_{\mathfrak{U}}$ as variables of a function $f : \mathfrak{X}_{\mathfrak{U}} \to \mathbb{R}$ is inconvenient since change of $\mathbf{X}$ is not totally free, we have to watch the conditions (1) and (2). To deal with this inconvenience and also to reduce the number of variables, we introduce another set of variables, the dihedral angles. First the definition and calculation of dihedral angles.

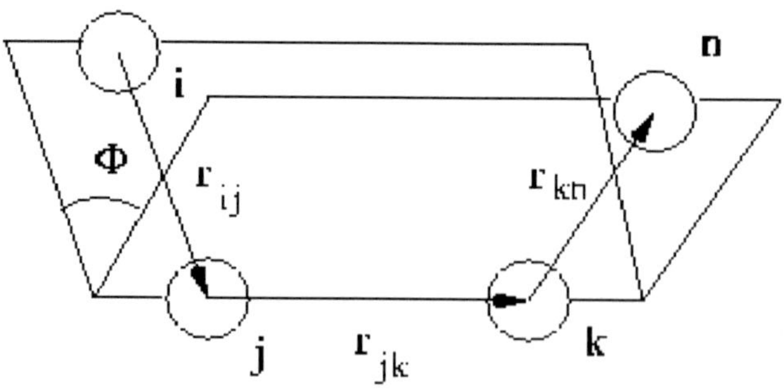

Figure 2. Definition of dihedral angle.

3.8.1 Dihedral Angles

Let $\wedge$ be the outer or cross product in $\mathbb{R}^3$. If in the molecule $\mathfrak{U}$, $\mathbf{a}_i$ has a bond with $\mathbf{a}_j$, $\mathbf{a}_j$ has a bond with $\mathbf{a}_k$, and $\mathbf{a}_k$ has a bond with $\mathbf{a}_n$, then it is possible that there will be a dihedral angle γ_{ijkn} in most conformations. Let $\mathbf{X} \in \mathfrak{X}_{\mathfrak{U}}$ be a conformation, let $\mathbf{r}_{ij} = \mathbf{x}_j - \mathbf{x}_i$, etc., if $\mathbf{x}_i, \mathbf{x}_j$, and $\mathbf{x}_k$ are not co-linear and $\mathbf{x}_j, \mathbf{x}_k$, and $\mathbf{x}_n$ are not co-linear, i.e., $\mathbf{r}_{ij} \wedge \mathbf{r}_{jk} \neq \vec{0}$, $\mathbf{r}_{jk} \wedge \mathbf{r}_{kn} \neq \vec{0}$, then they determine unique planes P_{ijk} and P_{jkn} respectively. In this case, there is a dihedral angle $\Phi = \gamma_{ijkn}$ in the conformation $\mathbf{X}$. It is the angle between the above two planes P_{ijk} and P_{jkn}. To calculate this angle in $[-\pi, \pi]$, we calculate unit normal vectors N_{ijk} and N_{jkn} of the planes P_{ijk} and P_{jkn} first. They are

$$N_{ijk} = \frac{\mathbf{r}_{ij} \wedge \mathbf{r}_{jk}}{|\mathbf{r}_{ij} \wedge \mathbf{r}_{jk}|}, \qquad N_{jkn} = \frac{\mathbf{r}_{jk} \wedge \mathbf{r}_{kn}}{|\mathbf{r}_{jk} \wedge \mathbf{r}_{kn}|}.$$

Note that they both are perpendicular to $\mathbf{r}_{jk}$. Thus, their cross product, $N_{ijk} \wedge N_{jkn}$, is a vector parallel to $\mathbf{r}_{jk}$ of the length $|\sin(\gamma_{ijkn})|$. Therefore,

$$N_{ijk} \wedge N_{jkn} = \frac{\mathbf{r}_{jk}}{|\mathbf{r}_{jk}|} \sin(\gamma_{ijkn}), \; \gamma_{ijkn} = \arcsin\left(\frac{\mathbf{r}_{jk} \bullet (N_{ijk} \wedge N_{jkn})}{|\mathbf{r}_{jk}|}\right). \quad (6)$$

Since the range of arcsin is $[-\pi, \pi]$, $\gamma_{ijkn} \in [-\pi, \pi]$.

In playing a transformer, we fix one part of the transformer and rotate another part, after a few such fixation-rotations, a transformer quickly changes its shape. Protein conformation in the form of $P_{\mathbf{X}}$ is just like a transformer, such a fixation-rotation changes dihedral angles. Before doing so, we have to do some preparation.

3.9 Changing of Dihedral Angles

3.9.1 Molecular Graphs

Molecules are comprised by atoms. These atoms are connected by chemical bonds, in biochemistry, they are usually covalent bonds. To describe bonding relations in a molecule, we will start with the general principle of molecular structures.

Given a molecule $\mathfrak{U}$, let V be the set of atoms in $\mathfrak{U}$ and $M = |V|$ be the number of atoms and label the atoms as $\mathbf{a}_1, \mathbf{a}_2, \cdots, \mathbf{a}_M$. For $1 \leq i, j \leq M$, define $B_{ij} = n$ if atoms $\mathbf{a}_i$ and $\mathbf{a}_j$ are connected by a bond with valency n (we can even imagine that n is not necessarily a whole number), if $\mathbf{a}_i$ and $\mathbf{a}_j$ do not form a bond, then $B_{ij} = 0$. The molecule formula of $\mathfrak{U}$ in chemistry can be seen as a graph $G(\mathfrak{U}) = (V, E)$, where V acts as the vertex set of $G(\mathfrak{U})$ and E is the edge set of $G(\mathfrak{U})$. An edge $\{i, j\} \in E$ if and only if $B_{ij} = n \geq 1$. We will call

$G(\mathfrak{U})$ the molecular graph of $\mathfrak{U}$. See FIGURE 3 and 4 for the molecular graphs of the amino acid Glycine and a benzene ring. Note that the double valency between C and O counts as only one edge.

In general, a graph $G = (V, E)$ allows that an edge $\{v, v\} \in E$. Such case can never happen in a molecular graph.

An important consequence of molecules living in the 3-dimensional space is that different molecules may have the same molecular graph and in that case only the different embeddings in the 3-dimensional space can distinguish them. For example, as shown in FIGURE 5, both the L- and D- forms of an amino acid have the same molecular graph. Therefore, molecular graphs cannot completely distinguish molecules.

But molecular graph can help us determine which dihedral angle is changeable.

3.9.2 Subgraphs

Let $G = (V, E)$ be a graph, a subgraph $G_i = (V_i, E_i)$ of G is a graph whose vertices form a subset of the vertex set V of G, $V_i \subset V$, and whose edges form a subset of E, $E_i \subset E$. $G_i \subset G$ is a full subgraph if all edges in G connecting vertices in G_i are also edges in G_i.

3.9.3 Paths and Connectivity

Let $G = (V, E)$ be a graph. A path in G from vertex v_1 to vertex v_n is a successive edges $(v_1, v_2), (v_2, v_3), \cdots, (v_{n-1}, v_n)$. G is connected if for any two vertices v and w we can find a path in G that goes from v to w.

When G is disconnected then there will be a set of full subgraphs G_i, such

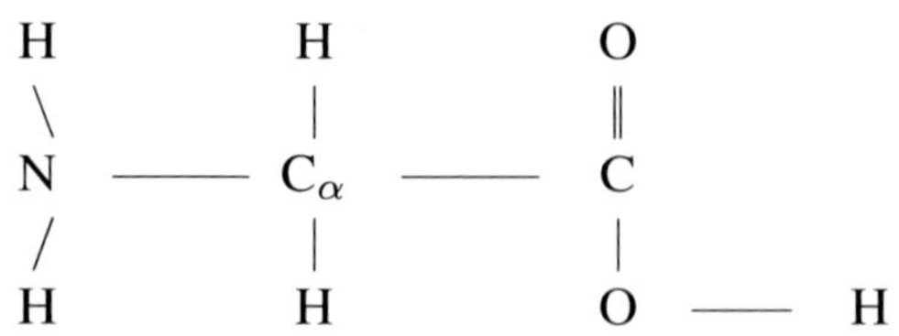

Figure 3. Glycine's molecular graph.

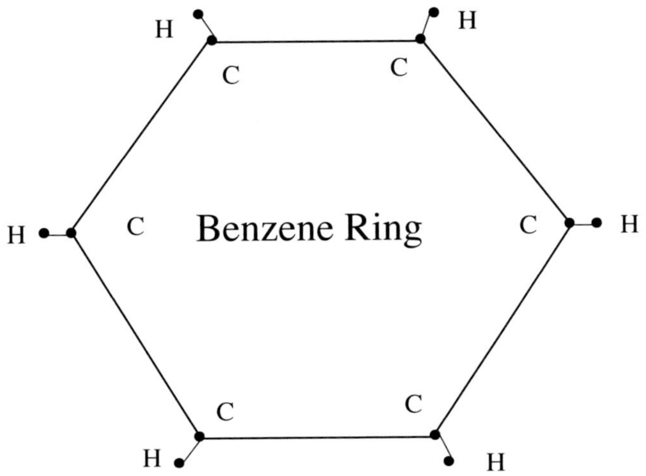

Figure 4. Loop in a Benzene's molecular graph.

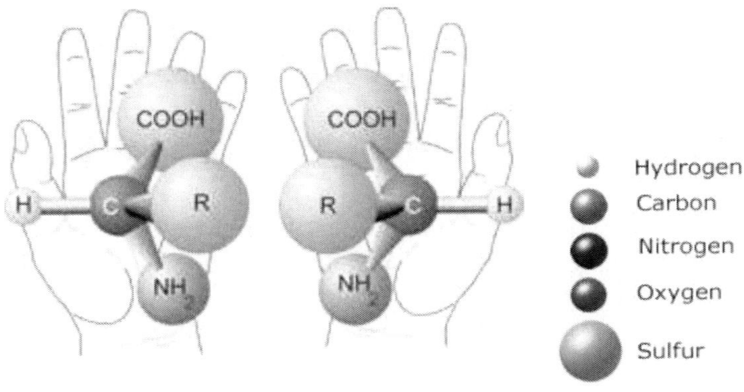

Figure 5. L- and D-forms of an amino acid and relative sizes and conventional colours of atoms in amino acids.

that $G = \cup_i G_i$, each G_i is connected and for any pair G_i and G_j, $i \neq j$, G_i and G_j have no common vertices, and for any two vertices $v_k \in G_i$ and $v_l \in G_j$

there are no paths in G that go from v_k to v_l. These G_is are called the connected components of G. If G is connected then it has only one connected component, itself.

Except for multi-chain proteins, all molecular graphs in this article are connected.

3.9.4 Loops and Simple Connectivity

A path $(v_1, v_2), (v_2, v_3), \cdots, (v_{n-1}, v_n)$ is a loop if $v_1 = v_n$ and there is at least one edge (v_i, v_{i+1}) in the path such that $v_i \neq v_{i+1}$. A connected graph G is simply connected if it does not contain any loop.

Since the benzene ring in FIGURE 4 of the benzene molecule $\mathfrak{U}$ forms a loop, $G(\mathfrak{U})$ is not simply connected, or it is non-trivial. One characteristic of a non-trivial graph is that if you cut off one edge from a loop of the graph, the remaining subgraph is still connected, i.e., for any two vertices in the remanning subgraph, there is a path connecting them in the remanning subgraph.

3.9.5 Rotatable Bonds

Let b_{ij} be a bond between $\mathbf{a}_i$ and $\mathbf{a}_j$ in $\mathfrak{U}$. In the molecular graph $G(\mathfrak{U})$ we cut off the edge corresponding to the bond to get a subgraph $G_{b_{ij}}$ of $G(\mathfrak{U})$. If $G_{b_{ij}}$ has exactly two connected components $G_1 \cup G_2 = G_{b_{ij}}$ (since $G(\mathfrak{U})$ is always connected, $G_{b_{ij}}$ can have at most two components) and each of them has at least two vertices, then b_{ij} is a rotatable bond.

If after cutting a bond b the remaining subgraph is connected, then b is not rotatable. For example, any bond in a Benzene ring is not a rotatable bond.

Covalent bonds of valency greater than 1 usually cannot be rotated, even though by our definition it may be rotatable. For example, all peptide bonds are not rotatable.

3.10 Transformer Rotations

3.10.1 Formula of Transformer Rotation

Let $\mathbf{b}_{ij} = \frac{\mathbf{x}_j - \mathbf{x}_i}{|\mathbf{x}_j - \mathbf{x}_i|}$ be the unit vector such that the rotation is counter-clockwise looking along the direction of the bond b_{ij}. Then for any point $X \in \mathbb{R}^3$, the rotation of $R_{ij\theta}$ is

$$R_{ij\theta}(X) = \mathbf{x}_i + [(X - \mathbf{x}_i)\bullet\mathbf{b}_{ij}]\mathbf{b}_{ij} + \cos\theta\{\mathbf{b}_{ij}\wedge[(X - \mathbf{x}_i)\wedge\mathbf{b}_{ij}]\} + \sin\theta[(X - \mathbf{x}_i)\wedge\mathbf{b}_{ij}]. \quad (7)$$

3.10.2 Transformer Rotation

Let b_{ij} be a rotatable bond. Let $V(G_k)$ be the set of vertices in G_k, $k = 1, 2$. Let $\mathbf{X} = (\mathbf{x}_1, \mathbf{x}_2, \cdots, \mathbf{x}_n)$ be a conformation. Let $R_{ij\theta}$ be a rotation of θ angle around the axis $\overline{\mathbf{x}_i\mathbf{x}_j}$ in $\mathbb{R}^3$. Define transformer rotation of $\mathbf{X}$ around $\overline{\mathbf{x}_i\mathbf{x}_j}$, $\mathfrak{R}_{ij\theta}$, as follows:

$$\mathfrak{R}_{ij\theta}(\mathbf{x}_k) = \begin{cases} R_{ij\theta}(\mathbf{x}_k), & \text{if } \mathbf{a}_k \in V(G_2); \\ \mathbf{x}_k, & \text{if } \mathbf{a}_k \in V(G_1). \end{cases}$$

Where $R_{ij\theta}$ is defined in (7).

The transformer rotated conformation $\mathbf{X}$, $\mathfrak{R}_{ij\theta}(\mathbf{X}) = (\mathfrak{R}_{ij\theta}(\mathbf{x}_1), \cdots, \mathfrak{R}_{ij\theta}(\mathbf{x}_n))$ will keep all the bond lengths and angles, and all dihedral angles except around $\overline{\mathbf{x}_i\mathbf{x}_j}$, as the same of those in $\mathbf{X}$. To see this, first note that a rotation in $\mathbb{R}^3$ does not change angles and lengths. Next for any two atoms of $\mathfrak{U}$ other than $\mathbf{a}_i$ and $\mathbf{a}_j$, if they form a covalent bond, then both of them are in either $V(G_1)$ or $V(G_2)$, because G_1 and G_2 are not connected in the subgraph $G_{b_{ij}}$. Thus both of them are either fixed or rotated. Finally, the following formula shows that $\mathfrak{R}_{ij\theta}(\mathbf{x}_k) = \mathbf{x}_k$ for $k = i, j$. Thus, for any $\mathbf{X} \in \mathfrak{X}_{\mathfrak{U}}$, $[\mathfrak{R}_{ij\theta}(\mathbf{X})] = [\mathbf{X}]$, i.e., transformer rotations keep the Equi-AD classes.

3.10.3 All Rotatable Dihedral Angles of $\mathfrak{U}$

We can list all rotatable bonds of a protein molecule $\mathfrak{U}$ as $\mathbf{B} = \{b_1, b_2, \cdots, b_L\}$, where $L \geq 1$ and is certainly much smaller than n, the number of atoms in $\mathfrak{U}$. Note that, like the hydrophobic levels, $\mathbf{B}$ is also independent of conformations of $\mathfrak{U}$. .

Remark 3.3. *We will allow forming new covalent bonds during the folding. But all these new covalent bonds will not be deliberately preserved, i.e., they can also be broken down in our manipulating of conformations. So the above* $\mathbf{B}$ *does not contain any of them.*

3.10.4 Commutativity of Transformer Rotations

Let $\mathfrak{R}_{ij\theta_1}$ and $\mathfrak{R}_{kl\theta_2}$ be two transformer rotations and $\mathbf{X} = (\mathbf{x}_1, \cdots, \mathbf{x}_n)$ be a conformation of $\mathfrak{U}$, it is proved in [14] that for any $\mathbf{x}_\alpha$, $1 \leq \alpha \leq n$,

$$\mathfrak{R}_{ij\theta_1} \circ \mathfrak{R}_{kl\theta_2}(\mathbf{x}_\alpha) = \mathfrak{R}_{kl\theta_2} \circ \mathfrak{R}_{ij\theta_1}(\mathbf{x}_\alpha).$$

Thus

$$\mathfrak{R}_{ij\theta_1} \circ \mathfrak{R}_{kl\theta_2}(\mathbf{X}) = \mathfrak{R}_{kl\theta_2} \circ \mathfrak{R}_{ij\theta_2}(\mathbf{X}). \tag{8}$$

3.10.5 Full Transformer Rotations

Let $b_i \in \mathbf{B}$ be a rotatable bond θ_i an angle, $i = 1, \cdots, L$, denote the transformer rotation of θ_i around the rotatable bond b_i as $\mathfrak{R}_{i,\theta_i}$. We will perform transformer rotations around all rotatable bonds in $\mathbf{B}$, that is, let $\mathbf{X}$ be a conformation, let $\Theta = (\theta_1, \cdots, \theta_L)$,

$$\mathfrak{R}_\Theta(\mathbf{X}) = \mathfrak{R}_{1,\theta_1} \circ \mathfrak{R}_{2,\theta_2} \circ \cdots \circ \mathfrak{R}_{L,\theta_L}(\mathbf{X}).$$

By (8), the order of operations in $\mathfrak{R}_\Theta$ does not matter, $\mathfrak{R}_\Theta$ is well defined. Of course, for any $\mathbf{X} \in \mathfrak{X}_{\mathfrak{U}}$, $[\mathfrak{R}_\Theta(\mathbf{X})] = [\mathbf{X}]$.

3.10.6 Dihedral Angle Variables

There may be more than one dihedral angle around a bond b. Once for all, we select one around each $b_i \in \mathbf{B}$. The conformations $\mathbf{X}$ and $\mathfrak{R}_\Theta(\mathbf{X})$ will have dihedral angles

$$\Phi(\mathbf{X}) = (\phi_1, \cdots, \phi_L), \quad \Phi(\mathfrak{R}_\Theta(\mathbf{X})) = (\phi_1 + \theta_1, \cdots, \phi_L + \theta_L).$$

Let $\mathbf{X}$ and $\mathbf{Y}$ be in the same Equi-DA class, $\Phi(\mathbf{X}) = (\phi_1, \cdots, \phi_L)$, $\Phi(\mathbf{Y}) = (\psi_1, \cdots, \psi_L)$. Let $\Theta = (\psi_1 - \phi_1, \cdots, \psi_L - \phi_L)$, then $\mathfrak{R}_\Theta(\mathbf{X}) = \mathbf{Y}$. Thus, starting from one conformation $\mathbf{X}$, by full transformer rotations we can achieve any conformation of the Equi-AD class $[\mathbf{X}]$. On the other hand, if $\mathbf{Y} \neq \mathbf{X}$ and $\mathbf{Y} \in [\mathbf{X}]$, then $\Phi(\mathbf{X}) \neq \Phi(\mathbf{Y})$. Thus the mapping:

$$\Phi : [\mathbf{X}] \to \overbrace{[0, 2\pi) \times \cdots \times [0, 2\pi)}^{L},$$

is one-to-one and onto and there is an inverse mapping:

$$\mathbf{Y}: \overbrace{[0,2\pi)\times\cdots\times[0,2\pi)}^{L} \to [\mathbf{X}].$$

Thus any function defined on $\mathfrak{X}_{\mathfrak{U}}$ can use Φ as variables, i.e.,

$$f : \mathfrak{X}_{\mathfrak{U}} \to \mathbb{R}, \quad f_{[\mathbf{X}]} : \overbrace{[0,2\pi)\times\cdots\times[0,2\pi)}^{L} \to \mathbb{R}$$

such that

$$f_{[\mathbf{X}]}(\Phi) = f(\mathbf{Y}(\Phi)). \tag{9}$$

$f_{[\mathbf{X}]}$ is periodic. Let $K = (k_1, \cdots k_L)$, k_i be integers, $i = 1, \cdots, L$,

$$\Psi = (\psi_1, \cdots, \psi_L) = (\phi_1 + 2k_1\pi, \cdots, \phi_L + 2k_L\pi) = \Phi + 2\pi K,$$

then

$$f_{[\mathbf{X}]}(\Phi) = f_{[\mathbf{X}]}(\Psi),$$

we can extend the definition of $f_{[\mathbf{X}]}$ to the whole $\mathbb{R}^L$. Then if f is lower semi-continuous, on the compact set $\overbrace{[0,2\pi]\times\cdots\times[0,2\pi]}^{L}$ it alway has at least a minimiser $\mathbf{Y} \in [\mathbf{X}]$.

4. Thermodynamics and Quantum Statistical Mechanics

To derive the Gibbs free energy function $G(\mathbf{X}; \mathfrak{E}_N, \mathfrak{U})$, we need thermodynamics and quantum statistics mechanics.

4.1 Thermodynamics

4.1.1 Thermodynamics System

A thermodynamic system consists of a bounded region $\mathfrak{S} \subset \mathbb{R}^3$ and materials inside $\mathfrak{S}$. We can measure quantities such as temperature T, pressure P, volume $V = V(\mathfrak{S})$, and the number of the material particles, N_i, $= 1, \cdots, l$, assuming that there are l kinds of particles. The heat bath of $\mathfrak{S}$ is the environment surrounding $\mathfrak{S}$, we can consider it as everything out of $\mathfrak{S}$ in $\mathbb{R}^3$, i.e., everything in $\mathfrak{S}' = \mathbb{R}^3 \setminus \mathfrak{S}$. Physically the wall of the system consists of the boundary $\partial\mathfrak{S} = \partial\mathfrak{S}'$ with various physical properties. If material particles and energy cannot be exchanged between $\mathfrak{S}$ and $\mathfrak{S}'$ through the wall $\partial\mathfrak{S}$, $\mathfrak{S}$ is an isolated system. If through the wall energy can be changed but not material particles, $\mathfrak{S}$ is a closed system. If both material particles and energy can be exchanged through the wall, $\mathfrak{S}$ is an open system.

After a relaxing time (maybe very long) the system will enter a equilibrium state, i.e., the system will no longer change with time, and is independent of its history, i.e., no mater what conditions it was at the beginning, at the end all will end in the same equilibrium state.

In equilibrium there are also state functions such as internal energy E, entropy S, pressure P, volume V, and particle numbers N_i, $i = 1, \cdots, l$. They are all constants for one particular $\mathfrak{S}$ that is in equilibrium, but not independent to each other, i.e., there are functional relations between them, for example, denoting $N = (N_1, \cdots, N_l)$, then $E = E(S, V, N)$, etc.

When a system is in a changing process, but changed relatively slowly such as each step of the process can be considered being infinitesimally close to a stated of equilibrium, it is in a quasi-static process. In a quasi-static process the thermodynamic identities in the following subsection are still hold.

A quasi-static process should be very slowly changing, so slow that it is only an idealised process, i.e., in nature no such processes exist. But in thermodynamical studies, a rapidly rotating heat engine can be considered as a quasi-static process, though its state changes up to several hundred times per second. This is important because the protein folding process will be treated as a quasi-static process.

4.1.2 Thermodynamic Identities

For an open system $\mathfrak{S}$, the fundamental equation is

$$\mathrm{d}E = T\mathrm{d}S - P\mathrm{d}V + \sum_{i=1}^{l} \mu_i \mathrm{d}N_i, \tag{10}$$

where μ_i's are chemical potentials. Thus $E = E(S, V, N)$. Indeed, equation (10) means that

$$\frac{\partial E}{\partial S} = T, \quad \frac{\partial E}{\partial V} = -P, \quad \frac{\partial E}{\partial N_i} = \mu_i.$$

Moreover, E is homogenous, i.e., for any $\lambda > 0$, we have

$$\lambda E(S, V, N) = E(\lambda S, \lambda V, \lambda N).$$

That means that

$$\begin{aligned} E(S, V, N) &= \frac{\mathrm{d}[\lambda E(S, V, N)]}{\mathrm{d}\lambda} = \frac{\mathrm{d}}{\mathrm{d}\lambda} E(\lambda S, \lambda V, \lambda N)\Big|_{\lambda=1} \\ &= S\frac{\partial E}{\partial S} + V\frac{\partial E}{\partial V} + \sum_{i=1}^{l} N_i \frac{\partial E}{\partial N_i} \\ &= TS - PV + \sum_{i=1}^{l} \mu_i N_i. \end{aligned} \tag{11}$$

Definition

$$F = E - TS, \quad G = F + PV$$

as the Helmholtz free energy and the Gibbs free energy respectively. By (11),

$$G = F + PV = E + PV - TS = \sum_{i=1}^{l} \mu_i N_i. \tag{12}$$

Similarly, by (12), the thermodynamic potential,

$$\Omega = E - TS - \sum_{i=1}^{l} \mu_i N_i,$$

satisfying

$$\Omega = E - TS - \sum_{i=1} \mu_i N_i = F - G = -PV. \tag{13}$$

4.2 Quantum Statistical Mechanics

To get the analytical formulas of thermodynamics functions such as entropy, we need statistical mechanics. Since we need quantum mechanics to describe the states in protein folding, we need quantum statistical mechanics.

4.2.1 Quantum States

Suppose $\mathfrak{S}$ is a thermodynamic system. In quantum mechanics a state of $\mathfrak{S}$ is a vector $|\psi\rangle$ in a complex Hilbert space $\mathfrak{H}$. For any $c \in \mathbb{C}$, the inner product $\langle\phi|\psi\rangle \in \mathbb{C}$ of two vectors $|\phi\rangle$ and $|\psi\rangle$ satisfies

$$\langle\phi|c\psi\rangle = c\langle\phi|\psi\rangle, \quad \langle c\phi|\psi\rangle = c^*\langle\phi|\psi\rangle, .$$

where c^* is the conjugate of $c \in \mathbb{C}$. We always normalise $|\psi\rangle$, i.e., $\langle\psi|\psi\rangle = 1$.

A physical observable A, for example, energy, corresponds to a Hermitian operator $\hat{A}$, such that the physical value of A at the state $|\psi\rangle$ is

$$\langle\hat{A}\rangle = \langle\psi|\hat{A}|\psi\rangle.$$

For example, the Hamiltonian $\hat{H}$ corresponds to energy, so that the state $|\psi\rangle$ has the internal energy E, and

$$E = \langle\hat{H}\rangle = \langle\psi|\hat{H}|\psi\rangle.$$

A state $|\psi\rangle$ is not unique, since for any $\alpha \in \mathbb{R}$, $e^{i\alpha}|\psi\rangle$ represents exactly the same state, only in a different phase. Hence we use the project operator $\hat{D} = |\psi\rangle\langle\psi|$, defined as

$$\hat{D} : \mathfrak{H} \to \mathfrak{H}, \quad \hat{D}(|\phi\rangle) = |\psi\rangle\langle\psi|\phi\rangle = \langle\psi|\phi\rangle\ |\psi\rangle, \quad \forall\ |\phi\rangle \in \mathfrak{H}.$$

$\hat{D}$ is independent of $e^{i\alpha}$ since

$$\langle e^{i\alpha}\psi|\phi\rangle\ e^{i\alpha}|\psi\rangle = e^{-i\alpha}e^{i\alpha}\langle\psi|\phi\rangle\ |\psi\rangle = \langle\psi|\phi\rangle\ |\psi\rangle, \quad \forall\ |\phi\rangle \in \mathfrak{H}.$$

Let $|i\rangle$, $i = 1, 2, 3, \cdots$, be a basis of $\mathfrak{H}$, then $|\psi\rangle = \sum_{i=1}^{\infty} c_i|i\rangle$.

$$\hat{D} = |\psi\rangle\langle\psi| = \left(\sum_{i=1}^{\infty} c_i|i\rangle\right)\left(\sum_{j=1}^{\infty} c_j^*\langle j|\right) = \sum_{1\le i,j<\infty} c_i c_j^*|i\rangle\langle j|.$$

Let

$$\hat{B} = \sum_{1\le i,j<\infty} b_{ij}|i\rangle\langle j|$$

be a Hermitian operator, then the trace of an $\hat{B}$ is $\mathbf{Tr}\,\hat{B} = \sum_{i=1}^{\infty} b_{ii}$, the summation is always finite. Also, the trace is independent of the basis $|i\rangle$ and

$$\mathbf{Tr}\,\hat{A}\hat{B} = \mathbf{Tr}\,\hat{B}\hat{A}. \tag{14}$$

For simplicity, we will select $|i\rangle$ be a orthonormal basis, i.e.,

$$\langle i|j\rangle = \delta_{ij}, \quad \delta_{ii} = 1, \quad \delta_{ij} = 0 \text{ for } i \neq j.$$

Let $\hat{A}$ be the Hermitian operator comes from a physical observable A, then

$$\begin{aligned} \hat{A}\hat{D} &= \hat{A}|\psi\rangle\langle\psi| = \left(\sum_{1\le i,j<\infty} a_{ij}|i\rangle\langle j|\right)\left(\sum_{k=1}^{\infty} c_k|k\rangle\right)\left(\sum_{l=1}^{\infty} c_l^*\langle l|\right) \\ &= \left(\sum_{1\le i,j,k<\infty} c_k a_{ij}|i\rangle\langle j|k\rangle\right)\left(\sum_{l=1}^{\infty} c_l^*\langle l|\right) = \sum_{1\le i,\, l<\infty}\left(\sum_{j=1}^{\infty} c_l^* a_{ij} c_j\right)|i\rangle\langle l|. \end{aligned}$$

Thus

$$\mathbf{Tr}\,\hat{A}\hat{D} = \sum_{j=1}^{\infty} c_i^* a_{ij} c_j.$$

Now at the state $|\psi\rangle$ we have

$$\begin{aligned} \langle\hat{A}\rangle &= \langle\psi|\hat{A}|\psi\rangle = \left(\sum_{l=1}^{\infty} c_l^*\langle l|\right)\left(\sum_{1\le i,j<\infty} a_{ij}|i\rangle\langle j|\right)\left(\sum_{k=1}^{\infty} c_k|k\rangle\right) \\ &= \left(\sum_{l=1}^{\infty} c_l^*\langle l|\right)\left(\sum_{1\le i,j<\infty} c_j a_{ij}|i\rangle\right) \\ &= \sum_{1\le i,\, l<\infty}\left(\sum_{j=1}^{\infty} c_l^* a_{ij} c_j\right)\langle l|i\rangle = \sum_{j=1}^{\infty} c_i^* a_{ij} c_j = \mathbf{Tr}\,\hat{A}\hat{D}. \end{aligned} \tag{15}$$

Two linear operators $\hat{A}$ and $\hat{B}$ communicate if

$$[\hat{A}, \hat{B}] = \hat{A}\hat{B} - \hat{B}\hat{A} = 0.$$

The eigenstates of any Hermitian operator $\hat{A}$ consists of a orthonormal basis of $\mathfrak{H}$. If $A_1, \cdots, A_k$ is a set of physical observable such that any two $\hat{A}_i$ and $\hat{A}_j$ communicate, then they have common eigenstates. The most complete information we can get from $|\psi\rangle$ is that we know each of $\langle\hat{A}_i\rangle$.

4.2.2 Density Operators

The state $|\psi\rangle$ is a pure state, the study of pure state is in the category of quantum mechanics. In most cases, we do not know which state our system $\mathfrak{S}$ is in. We only know that it may be in different states, say $|\psi_n\rangle$, $n = 1, 2, 3, \cdots$, with some probability. So for the same system $\mathfrak{S}$ we have an ensemble of states, we are in the category of statistical mechanics now. A density operator in this case is

$$\hat{D} = \sum_{n=1}^{\infty} |\psi_n\rangle p_n \langle\psi_n|, \quad p_n \geq 0, \quad \mathbf{Tr}\,\hat{D} = \sum_{n=1}^{\infty} p_n = 1.$$

Let $\hat{D}_n = |\psi_n\rangle\langle\psi_n|$, then $\hat{D} = \sum_{n=1}^{\infty} p_n \hat{D}_n$. Then by (15) the average of a physical observable A in all these states is given by

$$\langle\hat{A}\rangle = \sum_{n=1}^{\infty} p_n \langle\psi_n|\hat{A}|\psi_n\rangle = \sum_{n=1}^{\infty} p_n\,\mathbf{Tr}\,\hat{A}\hat{D}_n = \mathbf{Tr}\,\hat{A}\sum_{n=1}^{\infty} p_n \hat{D}_n = \mathbf{Tr}\,\hat{A}\hat{D}.$$

4.2.3 Statistical Entropy

We can diagnose $\hat{D}$. Let $|m\rangle$, $m = 1, 2, \cdots$, be an orthonormal basis of $\mathfrak{H}$, then the density operator $\hat{D}$ will be

$$\hat{D} = \sum_{m=1}^{\infty} |m\rangle P_m \langle m|, \quad P_m \geq 0, \quad \mathbf{Tr}\,\hat{D} = \sum_{m=1}^{\infty} P_m = 1.$$

Formally, we can define

$$\ln\hat{D} = \sum_{m=1}^{\infty} |m\rangle \ln P_m \langle m|, \quad \text{then } \hat{D}\ln\hat{D} = \sum_{m=1}^{\infty} |m\rangle P_m \ln P_m \langle m|.$$

Now we define the statistical entropy of $\hat{D}$ as

$$S_{st}[\hat{D}] = -k_B \sum_{m=1}^{\infty} P_m \ln P_m = -k_B\,\mathbf{Tr}\,\hat{D}\ln\hat{D}.$$

What is the relation between $S_{st}[\hat{D}]$ and the thermodynamics entropy S? So far, nothing. Because there are infinite many choices of the probability P_m, hence there are infinite many choices of the density operators. Which $\hat{D}$ is the one that will connect the $S_{st}[\hat{D}]$ to thermodynamic entropy S?

4.2.4 The Boltzmann Distribution

The good news is that if for a set of physical observables $A_1, A_2, \cdots, A_k$ such that $\hat{A}_1, \hat{A}_2, \cdots \hat{A}_k$ are commutative to each other, then there is a Boltzmann distribution of the probabilities P_M, that defines a density operator $\hat{D}_B$, such that for any density operator $\hat{D}$, if

$$\mathbf{Tr}\, \hat{A}_i \hat{D} = \mathbf{Tr}\, \hat{A}_i \hat{D}_B, \quad i = 1, \cdots, k, \tag{16}$$

then

$$S_{st}[\hat{D}_B] \geq S_{st}[\hat{D}]. \tag{17}$$

In fact, it can be proved that ([22])

$$\hat{D}_B = \frac{1}{Z} \exp\left(\sum_{i=1}^{k} \lambda_i \hat{A}_i\right), \quad Z = \mathbf{Tr} \exp\left(\sum_{i=1}^{k} \lambda_i \hat{A}_i\right), \quad \mathbf{Tr}\, \hat{D}_B = 1. \tag{18}$$

Remark 4.1. *Inequality in (17) was mistaken in [5] as what should be maximised (or minimised for corresponding free energy) in following the thermodynamic principle of protein folding. The author of [5] thinks that directly minimising the Gibbs free energy function $G(\mathbf{X}; \mathfrak{E}_N, \mathfrak{U})$ according to the thermodynamics principle is falling into a trap.*

Moreover, when the system $\mathfrak{S}$ is in equilibrium, or in a stage of a quasi-static process, $S_{st}[\hat{D}_B]$ is just the thermodynamic entropy $S(\mathfrak{S})$ of $\mathfrak{S}$ in subsection §§4.1.

4.2.5 Grand Canonic Ensemble

The λ_i's in (18) are Lagrange multipliers in the process of proving the maximality of $S_{st}[\hat{D}_B]$ in (17) under the conditions in (16). It is known that for $k = 1$, $\hat{A}_1 = \hat{H}$, $\lambda_1 = -\beta = -1/k_B T$. The $\hat{D}_B$ then is the density operator of of a canonic ensemble. Now suppose that in $\mathfrak{S}$ there are l kinds of material particles. Let N_j be the number of the jth particles, then $\hat{N}_j$ is a Hermitian operator. Set $\hat{A}_1 = \hat{H}$, $\hat{A}_i = \hat{N}_{i-1}$, $i = 2, \cdots l+1$, then $[\hat{A}_i, \hat{A}_j] = 0$, for $1 \leq i < j \leq l+1$, we have the density operator $\hat{D}_B$ of a grand canonic ensemble,

$$\hat{D}_B = \frac{1}{Z} \exp\left[-\beta\left(\hat{H} - \sum_{i=1}^{l} \mu_i \hat{N}_i\right)\right],$$

where

$$Z = \mathbf{Tr} \exp\left[-\beta\left(\hat{H} - \sum_{i=1}^{l} \mu_i \hat{N}_i\right)\right],$$

and μ_i is the chemical potential of the ith kind of particles. In particular we obtain the thermodynamic entropy $S(\mathfrak{S}) = S_{st}[\hat{D}_B]$, by (14),

$$\begin{aligned} S(\mathfrak{S}) &= -k_B \mathbf{Tr}\, \hat{D}_B \ln \hat{D}_B = -k_B \mathbf{Tr}\, \hat{D}_B \left\{\left[-\beta(\hat{H} - \sum_{i=1}^{l} \mu_i \hat{N}_i)\right] - \ln Z\right\} \\ &= k_B \ln Z + \frac{1}{T}\left(\langle \hat{H} \rangle - \sum_{i=1}^{l} \mu_i \langle \hat{N}_i \rangle\right) \quad (\mathbf{Tr}\, \hat{D}_B = 1) \end{aligned}$$

Hence we obtain

$$-k_B T \ln Z = \langle \hat{H} \rangle - TS(\mathfrak{S}) - \sum_{i=1}^{l} \mu_i \langle N_i \rangle.$$

The correspondence with formulas in §§4.1 are:

$$\langle \hat{H} \rangle = E, \quad \langle \hat{N}_i \rangle = N_i.$$

So that we have

$$-k_B T \ln Z = E - TS - \sum_{i=1}^{l} \mu_i N_i.$$

Compare with (13), we get that

$$-k_B T \ln Z = \Omega = -PV, \quad Z = e^{\beta PV}, \quad G = \sum_{i=1}^{l} \mu_i N_i.$$

Hence the density operator induced by the Boltzmann distribution of a grand canonic ensemble on a thermodynamic system $\mathfrak{S}$ is given by

$$\hat{D}_B = \exp\left[-\beta\left(\hat{H} + PV(\mathfrak{S}) - \sum_{i=1}^{l} \mu_i \hat{N}_i\right)\right]. \tag{19}$$

5. Derivation of $G(\mathbf{X}; \mathfrak{E}_N, \mathfrak{U})$ for a Globular Protein $\mathfrak{U}$

Recall that our Gedanken experiment of watching and recording the loci of a single protein molecule folding. The loci is a family of conformations $\mathbf{X}(t)$, $0 \le t \le t_d$. In a crowed cell, although separated from other molecules of the same protein, a protein molecule is always surrounded by other molecules and ions, etc.. And suppose we have a tiny thermodynamic system $\mathfrak{S}_t$ for each $\mathbf{X}(t)$ that contains not only $P_{\mathbf{X}(t)}$ but also other material particles. The thermodynamic system $\mathfrak{S}_t$ is so tiny that direct observation is impossible, but luckily in thermodynamics there is no size restrictions in the definition of thermodynamic system, hence we can construct such $\mathfrak{S}_t$ mentally anyway. After that, thermodynamics and quantum statistical mechanics in Section §4 really can be applied to get the Gibbs free energy function $G(t) = G(\mathfrak{S}_t)$. That is what we are going to do.

5.1 The Thermodynamic System $\mathfrak{S}_\mathbf{X}$ Associated with a Conformation $\mathbf{X}$

Suppose $\mathbf{X} = (\mathbf{x}_1, \cdots, \mathbf{x}_n) \in \mathfrak{X}_\mathfrak{U}$. Recall the $B(\mathbf{x}_i)$ in subsection §§3.5 and $P_\mathbf{X}$ in (5). The conformation $\mathbf{X}$ actually occupies a space in $\mathbb{R}^3$ that is a little larger than $P_\mathbf{X}$, we will write it as $\Omega_\mathbf{X} \subset \mathbb{R}^3$, with boundary $\Sigma_\mathbf{X} = \partial\Omega_\mathbf{X}$. The $\Omega_\mathbf{X}$ is a bounded domain in $\mathbb{R}^3$ such that $P_\mathbf{X} \subset \overline{\Omega}_\mathbf{X}$ and no room left for other material particles inside $\overline{\Omega}_\mathbf{X}$. Where $\overline{\Omega}_\mathbf{X} = \Omega_\mathbf{X} \cup \Sigma_\mathbf{X}$ is the closure of $\Omega_\mathbf{X}$, i.e., the smallest closed set containing $\Omega_\mathbf{X}$. If we denote $\Omega'_\mathbf{X} = \mathbb{R}^3 \backslash \overline{\Omega}_\mathbf{X}$, then

$$\Sigma_\mathbf{X} = \partial\Omega_\mathbf{X} = \partial\Omega'_\mathbf{X}; \quad \mathbb{R}^3 = \Omega_\mathbf{X} \cup \Sigma_\mathbf{X} \cup \Omega'_\mathbf{X}.$$

Recall that the distance between two points $\mathbf{x} = (x_1, x_2, x_3) \in \mathbb{R}^3$ and $\mathbf{y} = (y_1, y_2, y_3) \in \mathbb{R}^3$ is $|\mathbf{x} - \mathbf{y}| = \sqrt{\sum_{i=1}^3 (x_i - y_i)^2}$. Define the distance of a point $\mathbf{x}$ to a subset $S \subset \mathbb{R}^3$ by $\mathrm{dist}(\mathbf{x}, S) = \inf_{\mathbf{y} \in S} |\mathbf{x} - \mathbf{y}|$.

Let $d = d_\mathfrak{E} > 0$, where $d_\mathfrak{E}$ is the largest diameter of material particles in the environment $\mathfrak{E}$, define

$$\mathcal{R}^d_\mathbf{X} = \{\mathbf{x} \in \Omega'_\mathbf{X} : \ \mathrm{dist}(\mathbf{x}, \Sigma_\mathbf{X}) \le d\}, \quad \mathfrak{S}^d_\mathbf{X} = \Omega_\mathbf{X} \cup \mathcal{R}^d_\mathbf{X}.$$

The $\mathfrak{S}^d_\mathbf{X}$ is our thermodynamic system tailor-made for the conformation $\mathbf{X}$. Besides $P_\mathbf{X}$, it contains only one layer of material particles of the immediate environment of $P_\mathbf{X}$. Note that such a $\mathfrak{S}^d_\mathbf{X}$ is necessarily very small, for example, if the material particles are water molecules and make $d = d_w$, where d_wÅ is the diameter of a water molecule, then for a not too extended conformation $\mathbf{X} \in \mathfrak{X}_{\mathfrak{U}}$ of a protein molecule $\mathfrak{U}$ which has several hundred residues, the diameter of $\mathfrak{S}^{d_w}_{\mathbf{X}}$ will be only several nanometers. Such a small system is beyond reaches of any current precise measurement in biophysics.

Note that in $\mathfrak{S}^d_{\mathbf{X}}$, there is only one protein molecule, and that molecule is in the three dimensional shape $P_{\mathbf{X}}$. Hence for different conformations $\mathbf{X}$ and $\mathbf{Y}$, $\mathfrak{S}^d_{\mathbf{X}} \neq \mathfrak{S}^d_{\mathbf{Y}}$.

Furthermore, $\mathfrak{S}^d_{\mathbf{X}}$ not only depends on conformation $\mathbf{X}$, but also depends on environment $\mathfrak{E}$, via $d = d_{\mathfrak{E}}$, the largest diameter of material particles in $\mathfrak{E}$. Hence it should be denoted as $\mathfrak{S}^{d_{\mathfrak{E}}}_{\mathbf{X},\mathfrak{E}}$. Again, for different environments $\mathfrak{E}_1$ and $\mathfrak{E}_2$, generally $\mathfrak{S}^{d_{\mathfrak{E}_1}}_{\mathbf{X},\mathfrak{E}_1} \neq \mathfrak{S}^{d_{\mathfrak{E}_2}}_{\mathbf{X},\mathfrak{E}_2}$, unless the material particles of $\mathfrak{E}_1$ and $\mathfrak{E}_2$ are the same and $\mathfrak{E}_1$ and $\mathfrak{E}_2$ in different thermodynamics states, for example, they differ only in temperature T, pressure P, etc.

Since we will concentrate on the physiological environment $\mathfrak{E}_N$, and among all proteins, globular proteins have the simplest physiologic environment $\mathfrak{E}_N$, essentially there are only water molecules in $\mathfrak{E}_N$. Therefore, from now on, we will simply write $\mathfrak{S}_{\mathbf{X}}$ instead of $\mathfrak{S}^{d_{\mathfrak{E}_N}}_{\mathbf{X},\mathfrak{E}_N}$, where $d_{\mathfrak{E}_N} = d_w$. Accordingly, we write $\mathcal{R}_{\mathbf{X}}$ instead of $\mathcal{R}^{d_w}_{\mathbf{X}}$.

5.1.1 Partitioning of $\mathcal{R}_{\mathbf{X}}$

Recall the hydrophobicity classes $\cup_{i=1}^k H_i = \{\mathbf{a}_1, \cdots, \mathbf{a}_n\}$. Let $B(\mathbf{x}_i)$ be the atomic space in §§3.5 and $P_{\mathbf{X}}$ in (5), we define

$$P_{\mathbf{X},i} = \cup_{\mathbf{a}_j \in H_i} B(\mathbf{x}_j), \quad i = 1, \cdots, k, \quad P_{\mathbf{X}} = \cup_{i=1}^k P_{\mathbf{X},i} \subset \Omega_{\mathbf{X}}.$$

For a fixed conformation $\mathbf{X}$, according to the hydrophobic levels of atoms, we can divide the material layer $\mathcal{R}_{\mathbf{X}}$. Define

$$\mathcal{R}_{\mathbf{X},i} = \{\mathbf{x} \in \mathcal{R}_{\mathbf{X}} : \ \mathrm{dist}(\mathbf{x}; P_{\mathbf{X},i}) \leq \mathrm{dist}(\mathbf{x}, P_{\mathbf{X}} \setminus P_{\mathbf{X},i})\}.$$

Recall that $\Sigma_{\mathbf{X}} = \partial\Omega_{\mathbf{X}}$ is the boundary of $\Omega_{\mathbf{X}}$. Let

$$\Sigma_{\mathbf{X},i} = \Sigma_{\mathbf{X}} \cap \mathcal{R}_{\mathbf{X},i}, \quad i = 1, \cdots, k.$$

We will label H_i such that H_1 is the strongest hydrophobic and H_k is the strongest hydrophilic. There is a $l < k$ such that H_i is hydrophobic for $i \leq l$, and H_i is hydrophilic for $i > l$. There will be chemical potentials (they are Gibbs free energy for each individual water molecule in $\mathcal{R}_{\mathbf{X},i}$, $i = 1, \cdots, k$.

$$\mu_1 > \mu_2 > \cdots > \mu_l > 0 > \mu_{l+1} > \cdots > \mu_k.$$

Similarly, the Gibbs free energy per electron in $\mathfrak{S}_\mathbf{X}$ is the chemical potential μ_e.

Then define the hydrophobic and hydrophilic surfaces, $\Sigma_{\mathbf{X},h}$ and $\Sigma_{\mathbf{X},p}$ of the conformation $\mathbf{X}$ as

$$\Sigma_{\mathbf{X},h} = \cup_{i=1}^{l} \Sigma_{\mathbf{X},i}, \quad \Sigma_{\mathbf{X},p} = \cup_{i=l+1}^{k} \Sigma_{\mathbf{X},i} = \Sigma_\mathbf{X} \backslash \Sigma_{\mathbf{X},h}. \tag{20}$$

5.1.2 Open Systems and Grand Canonic Ensembles

Let $\mathfrak{U}$ be a globular protein and $\mathbf{X} \in \mathfrak{X}_\mathfrak{U}$. Inside $\mathfrak{S}_\mathbf{X}$, not only there is $P_\mathbf{X}$, but also other material particles. To simplify the system, we will only allow absolutely essential particles to go into $\mathfrak{S}_\mathbf{X}$. For example, some globular proteins not only fold in a cell, but also can fold out of a cell in solvents such as water. So that the essential particles in $\mathfrak{S}_\mathbf{X}$, in case that $\mathfrak{U}$ is globular, are only water molecules. In this sense, globular proteins have the simplest environment. Other kinds of proteins will have more complicated environments, for example, membrane proteins have to deal with hydrophobic membrane environment and environments inside and outside of the cell membrane. Because of the simplicity of their environment, in this article we will only consider globular proteins.

Suppose in a state there are N_e electrons in $\mathfrak{S}_\mathbf{X}$ and N_i water molecules in $\mathcal{R}_{\mathbf{X},i}$. Let $N = (N_e, N_1, \cdots, N_k)$. We will consider all cases of N, as long as each $\mathcal{R}_{\mathbf{X},i}$ can contain N_i water molecules and $\mathfrak{S}_\mathbf{X}$ can contain N_e electrons. In another word, electrons and water molecules are allowed to enter and leave $\mathfrak{S}_\mathbf{X}$ and $\mathcal{R}_\mathbf{X}$ respectively. Such an thermodynamic system is an open system. We further require the system has fixed temperature and pressure. The ensemble of electrons and water molecules and their quantum states consists of a grand canonic ensemble.

5.2 The Hamiltonian

Let m_i be the mass of the atom $\mathbf{a}_i$ in $\mathfrak{U}$, m_e the mass of an electron, and m_w the mass of a water molecule. We will treat a water molecule as a particle centred at the atomic centre of its oxygen atom. Let $L_N = \sum_{i=1}^{k} N_i$.

For each $N = (N_e, N_1, \cdots, N_k)$, the Hamiltonian of the system $\mathfrak{S}_\mathbf{X}$ is that

$$\hat{H}_N = \hat{T}_N + \hat{V}_N,$$

where

$$\hat{T}_N = -\sum_{i=1}^{n} \frac{\hbar}{2m_i}\nabla_i - \frac{\hbar}{2m_w}\sum_{i=1}^{L_N}\nabla_i - \frac{\hbar}{2m_e}\sum_{i=1}^{N_e}\nabla_i.$$

The usual Coulombs potential energy will be

$$V_N = V_{\mathbf{X}} + V_{N,\mathbf{X}}.$$

Let Z_i be the number of protons in $\mathbf{a}_i$, then $V_{\mathbf{X}}$ is the Coulombs potential

$$V_{\mathbf{X}} = \sum_{1 \le i < j \le n} \frac{Z_i Z_j}{|\mathbf{x}_i - \mathbf{x}_j|}.$$

It is independent of water molecules and electrons, hence is independent of N.

5.2.1 Adiabatic Approximation

Now because that $\mathfrak{U}$ has much bigger mass than any particles in $\mathfrak{S}_{\mathbf{X}}$, we will fix the conformation $\mathbf{X}$, and let all particles in $\mathcal{R}_{\mathbf{X}}$ and electrons in $\mathfrak{S}_{\mathbf{X}}$ move. The corresponding Hamiltonian then is

$$\hat{H}_{N,\mathbf{X}} = -\frac{\hbar}{2m_w}\sum_{i=1}^{L_N}\nabla_i - \frac{\hbar}{2m_e}\sum_{i=1}^{N_e}\nabla_i + \hat{V}_{N,\mathbf{X}}. \tag{21}$$

Remark 5.1. $\hat{H}_{N,\mathbf{X}}$ *in* (21) *is an extension of the Born-Oppenheimer approximation. See [6] and [18]. The potential energy part in* $\hat{H}_{N,\mathbf{X}}$ *is slightly different between them, we follow [18].*

Let $\hat{N}_i(\mathbf{X})$ be the Hermitian operators and $\mathbb{N}$ be the set of nature numbers. The conformation $\mathbf{X}$ then theoretically will have all energy levels $E^i_{N,\mathbf{X}}$ (eigenvalues of the $\hat{H}_{N,\mathbf{X}}$), i.e., there are eigenstates $\Psi_i^{N,\mathbf{X}}$ such that

$$\hat{H}_{N,\mathbf{X}}\Psi_i^{N,\mathbf{X}} = E^i_{N,\mathbf{X}}\Psi_i^{N,\mathbf{X}}, \quad \hat{N}_i(\mathbf{X})\Psi_i^{N,\mathbf{X}} = N_i\Psi^{N,\mathbf{X}}, \quad i \in \mathbb{N}. \tag{22}$$

The Hilbert space $\mathfrak{H}_N$ is $L_0^2(\mathfrak{S}_{\mathbf{X}}^{N_e} \times \prod_{i=1}^k \mathcal{R}_{\mathbf{X},i}^{N_i})$. Finally define

$$\mathfrak{H} = \oplus_N \mathfrak{H}_N, \quad \hat{H}_{\mathbf{X}} = \oplus_N \hat{H}_{\mathbf{X},N}.$$

5.3 Numerical Version of $G(\mathbf{X}; \mathfrak{C}_N, \mathfrak{U})$

Let $V(\mathbf{X}) = V(\mathfrak{S}_\mathbf{X})$ be the volume of $\mathfrak{S}_\mathbf{X}$ and $S(\mathbf{X}) = S(\mathfrak{S}_\mathbf{X})$ the entropy of $\mathfrak{S}_\mathbf{X}$. By (19), we have the density operator $\hat{D}_B(\mathbf{X})$ of the open thermodynamic system $\mathfrak{S}_\mathbf{X}$,

$$\hat{D}_B(\mathbf{X}) = \exp[-\beta(\hat{H}_\mathbf{X} + PV(\mathbf{X}) - \mu_e\hat{N}_e - \sum_{i=1}^{k}\mu_i\hat{N}_i)],$$

Therefore, by (13)

$$\begin{aligned} S(\mathbf{X}) &= -k_B\,\mathbf{Tr}\,\hat{D}_B(\mathbf{X})\ln\hat{D}_B(\mathbf{X}) \\ &= -k_B\,\mathbf{Tr}\,\hat{D}_B(\mathbf{X})\left[-\beta\left(\hat{H}_\mathbf{X} + PV(\mathbf{X}) - \mu_e\hat{N}_e - \sum_{i=1}^{k}\mu_i\hat{N}_i\right)\right] \\ &= \frac{1}{T}\left(\langle\hat{H}_\mathbf{X}\rangle + PV(\mathbf{X}) - \mu_e\langle N_e\rangle - \sum_{i=1}^{k}\mu_i\langle N_i\rangle\right), \end{aligned}$$

and

$$\langle\hat{H}_\mathbf{X}\rangle + PV(\mathbf{X}) - TS(\mathbf{X}) = \mu_e\langle N_e\rangle + \sum_{i=1}^{k}\mu_i\langle N_i. \tag{23}$$

Usually $\langle\hat{H}_\mathbf{X}\rangle$ should be the internal energy $E(\mathfrak{S}_\mathbf{X})$ of $\mathfrak{S}_\mathbf{X}$. By definition $E(\mathfrak{S}_\mathbf{X})$ should be the total sum of all kinetic and potential energies. In our case, we have left the potential energy $V_\mathbf{X}$ out of the Hamiltonian. Therefore, to count the internal energy, we have

$$E(\mathbf{X}) = E(\mathfrak{S}_\mathbf{X}) = \langle\hat{H}_\mathbf{X}\rangle + V_\mathbf{X}, \tag{24}$$

thus by (12), (23), and (24) we obtain the Gibbs free energy

$$\begin{aligned} G(\mathbf{X}; \mathfrak{C}_N, \mathfrak{U}) &= G(\mathfrak{S}_X) = E(\mathbf{X}) + PV(\mathbf{X}) - TS(\mathbf{X}) \\ &= \mu_e\langle\hat{N}_e\rangle + \sum_{i=1}^{k}\mu_i\langle\hat{N}_i\rangle + V_\mathbf{X} \\ &= \mu_e\langle\hat{N}_e\rangle + \sum_{i=1}^{k}\mu_i\langle\hat{N}_i\rangle + \sum_{1\le i<j\le n}\frac{Z_iZ_j}{|\mathbf{x}_i - \mathbf{x}_j|}. \end{aligned} \tag{25}$$

5.4 Geometric Version of $G(\mathbf{X}; \mathfrak{E}_N, \mathfrak{U})$

Formula (25) will be useful if we can really count $\langle \hat{N}_e \rangle$ and $\langle \hat{N}_i \rangle$. This needs a good understanding of the geometry of $\Omega_{\mathbf{X}}$ and $\mathcal{R}_{\mathbf{X},i}$ to estimate the maximum water molecules can be contained in $\mathcal{R}_{\mathbf{X},i}$ (that should be $\langle \hat{N}_i \rangle$ simply because nature will not leave large void).

A simple geometric explanation of (25) will give us an geometric version of $G(\mathbf{X}; \mathfrak{E}_N, \mathfrak{U})$, via the volume of $\Omega_{\mathbf{X}}$ and areas of $\Sigma_{\mathbf{X}}$ and $\Sigma_{\mathbf{X},i}$.

Since every water molecule in $\mathcal{R}_{X,i}$ has contact with the surface $\Sigma_{X,i}$, $\langle \hat{N}_i \rangle$ is proportional to the area $A(\Sigma_{X,i})$. Therefore, there are $\nu_i > 0$, such that

$$\nu_i A(\Sigma_{X,i}) = \langle \hat{N}_i(X) \rangle, \quad 1 \leq i \leq k. \tag{26}$$

Let $p_{\mathbf{X},\mathbf{N}}(\mathbf{x})$ be the electronic density distribution for $N = (N_e, N_1, \cdots, N_k)$. By [3],

$$N_e = \int_{\mathfrak{S}_{\mathbf{X}}} p_{\mathbf{X},\mathbf{N}}(\mathbf{x}) d\mathbf{x}.$$

The mean value theorem for integrals guarantees that there is a $\nu_{\mathbf{X},N} > 0$ such that

$$N_e = \int_{\mathfrak{S}_{\mathbf{X}}} p_{\mathbf{X},N}(\mathbf{x}) d\mathbf{x} = \nu_{\mathbf{X},N} V(\mathfrak{S}_{\mathbf{X}}).$$

By the definition of $\mathfrak{S}_{\mathbf{X}}$ and $\Omega_{\mathbf{X}}$, we have roughly $V(\mathfrak{S}_{\mathbf{X}} \setminus \Omega_{\mathbf{X}}) = d_w A(\Sigma_{\mathbf{X}})$. Then taking the mean we have

$$\begin{aligned} \langle \hat{N}_e(\mathbf{X}) \rangle &= \langle \hat{\nu}_{\mathbf{X},N} V(\mathfrak{S}_{\mathbf{X}}) \rangle = \langle \hat{\nu}_{\mathbf{X},N} \rangle V(\mathfrak{S}_{\mathbf{X}}) \\ &= \nu_e V(\mathfrak{S}_{\mathbf{X}}) = \nu_e [V(\Omega_{\mathbf{X}}) + V(\mathfrak{S}_{\mathbf{X}} \setminus \Omega_{\mathbf{X}})] \\ &= \nu_e V(\Omega_{\mathbf{X}}) + \nu_e d_w A(\Sigma_{\mathbf{X}}). \end{aligned} \tag{27}$$

Substitute (26) and (27) into (25), letting $\omega_e = \nu_e \mu_e$, $\omega_i = \nu_i \mu_i$, we obtain

$$G(\mathbf{X}; \mathfrak{E}_N, \mathfrak{U}) = \omega_e V(\Omega_{\mathbf{X}}) + \omega_e d_w A(\Sigma_{\mathbf{X}}) + \sum_{i=1}^{k} \omega_i A(\Sigma_{\mathbf{X},i}) + \sum_{1 \leq i < j}^{n} \frac{Z_i Z_j}{|\mathbf{x}_i - \mathbf{x}_j|}. \tag{28}$$

Remark 5.2. *Our derivation of the Gibbs free energy function $G(\mathbf{X}; \mathfrak{E}_N, \mathfrak{U})$ is, of course via statistical mechanics. The difference is that we do not follow the influence of ensemble measurement practiced in experiments. That influence made people to consider an ensemble of the same protein molecule,*

$(\mathbf{X}_1, \mathbf{X}_2, \cdots, \mathbf{X}_N)$, with large N. Instead, we consider only one protein molecule $\mathfrak{U}$ and consider an arbitrary conformation of it, $\mathbf{X}$. To apply statistical mechanics, we consider a tailor-made thermodynamic system $\mathfrak{S}_{\mathbf{X}}$, and an ensemble of other molecules or articles inside it. The system $\mathfrak{S}_{\mathbf{X}}$ will be an open system thus we will apply the grand canonic ensemble to particles except to the protein molecule $\mathfrak{U}$ at the conformation $\mathbf{X}$. Thus equation (23) give us the Gibbs free energy of the system $\mathfrak{S}_{\mathbf{X}}$. To translate it into an analytic formula, we need combine the geometric expression of the conformation $\mathbf{X}$ and the special property of the proteins, they are amphipathic.

Once we get the Gibbs free energy function $G(\mathbf{X}; \mathfrak{C}_N, \mathfrak{U})$, quantum statistical mechanics as a tool in our protein folding study stops here. The remaining is applying $G(\mathbf{X}; \mathfrak{C}_N, \mathfrak{U})$ to reveal the secret of protein folding.

Remark 5.3. *Since 1950s hydrophobic effect has been regarded as the dominate force of protein folding, [21] and [10]. Another view is that it is not the hydrophobic effect, but the hydrophilic interactions that push the protein to fold, [5]. No matter it is hydrophobic effect or hydrophilic force, the observed fact is that globular proteins have a well formed hydrophobic core in its native structure, i.e., hydrophobic residues tend to hide inside the native structure, avoid being exposed to the boundary surface of the structure, therefore, less contacts with water. Furthermore, native structures also have smaller volume and surface area than non-native structures, [29] and [30]. The derived Gibbs free energy formula in* (28) *expresses both hydrophobic effect and hydrophilic force, indeed, since $\mu_i > 0$ for a hydrophobic level, smaller area $A(\Sigma_{\mathbf{X},i})$ means smaller Gibbs free energy; for hydrophilic level H_j, $j > l$, $\mu_j < 0$, hence, bigger area $A(\Sigma_{\mathbf{X}j})$ means smaller Gibbs free energy too. So these hydrophobic level surface term explains the root of hydrophobic core: hydrophobic moieties tend to hide inside and hydrophilic moieties tend to expose to the surface, the result is smaller Gibbs free energy. Also the terms of volume and area, because their positive coefficients, shrinking volume $V(\Omega_{\mathbf{X}})$ and area $A(\Sigma_{\mathbf{X}})$ also contributes to reducing the Gibbs free energy. This combination of hydrophobic effect and hydrophilic force may stop the arguments about which one is the driving force of protein folding. Actually, they all are.*

Remark 5.4. *Based on above mentioned observations made in [29] and [30], a phenomenological model of protein folding was formed by intentionally shrinking the volume and surface area of $\Omega_{\mathbf{X}}$, and make the conformation have better hydrophobic core, [13]. To express the hydrophobic core, hydrophobic surface* *was invented, the definition is just like that in* (20)*, only simpler,* $k = 2$*, there are just two levels of hydrophobicity, hydrophobic or hydrophilic. Indeed, if* $k = 2$*, writing* $\omega_1 = \omega_h$ *and* $\omega_2 = \omega_p$*, then*

$$A(\Sigma_{\mathbf{X},1}) = A(\Sigma_{\mathbf{X},p}) = A(\Sigma_{\mathbf{X}}) - A(\Sigma_{\mathbf{X},h}) = A(\Sigma_{\mathbf{X}}) - A(\Sigma_{\mathbf{X},2}),$$

(28) *becomes*

$$\begin{aligned} G(\mathbf{X}; \mathfrak{E}_N, \mathfrak{U}) &= \omega_e V(\Omega_{\mathbf{X}}) + (\omega_e d_w + \omega_p) A(\Sigma_{\mathbf{X}}) \\ &\quad + (\omega_h - \omega_p) A(\Sigma_{\mathbf{X},h}) + \sum_{1 \le i < j \le n}^{n} \frac{Z_i Z_j}{|\mathbf{x}_i - \mathbf{x}_j|}. \end{aligned}$$

Except the last term, it is essentially the mathematical model in [13].

6. Mathematical Properties of $G(\mathbf{X}; \mathfrak{E}_N, \mathfrak{U})$

The mathematical property of the Gibbs free energy function is important when we apply the function to concrete problems in protein folding. So we will collect these useful fact together in one section.

6.1 Congruence Invariant

Let $C : \mathbb{R}^3 \to \mathbb{R}^3$ be a congruence, i.e., parallel translations, rotations, and any composition of them. Defined $C(\mathbf{X}) = (C(\mathbf{x}_1), \cdots, C(\mathbf{x}_n))$. As conformations $\mathbf{X}$ and $C(\mathbf{X})$ will be the same because all relative relations such as the distance between the $\mathbf{x}_i$'s, bond lengths and angles, are kept in $C(\mathbf{X})$ as exactly the same. We distinguish $\mathbf{X}$ and $C(\mathbf{X})$ as different element in $\mathfrak{X}_{\mathfrak{U}}$ because that we want to use the inherited topology on $\mathfrak{X}_{\mathfrak{U}} \subset \mathbb{R}^{3n}$. But then any function f defined on conformations, if it is defined as $\mathfrak{X}_{\mathfrak{U}}$, has to satisfy the condition of congruence invariant, i.e., f should satisfy

$$f(\mathbf{X}) = f(C(\mathbf{X})), \quad \text{for any congruence } C : \mathbb{R}^3 \to \mathbb{R}^3. \tag{29}$$

Our Gibbs free energy function in (25) and (28) should not be exception.

Theorem 6.1. *The Gibbs free energy function in both the numerical version and the geometric version are congruence invariant.*

Proof. Let $\mathbf{X} = (\mathbf{x}_1, \cdots, \mathbf{x}_n)$ be a conformation and C a congruence. Remember that in derivation of $G(\mathbf{X}; \mathfrak{E}_N, \mathfrak{U})$ we first form $P_{\mathbf{X}} = \cup_{i=1}^n B(\mathbf{x}_i)$, then form $\mathcal{R}_{\mathbf{X}}$, $\Omega_{\mathbf{X}}$, and $\mathfrak{S}_{\mathbf{X}} = \Omega_{\mathbf{X}} \cup \mathcal{R}_{\mathbf{X}}$. Since $C(P_{\mathbf{X}}) = \cup_{i=1}^n C(\mathbf{x}_i)) = P_{C(\mathbf{X})}$, etc., eventually we have $C(\mathfrak{S}_{\mathbf{X}}) = \mathfrak{S}_{C(\mathbf{X})}$. Since the hydrophobicity classes are independent of conformations, it is also true that $C(\mathcal{R}_{\mathbf{X},i}) = \mathcal{R}_{C(\mathbf{X}),i}$.

Since volume, area, and distance are invariant under a congruence, the geometric version of $G(\mathbf{X}; \mathfrak{E}_N, \mathfrak{U})$ is congruence invariant. Since volume, area, distance together decide how many material particle can be in a space, the numerical version of $G(\mathbf{X}; \mathfrak{E}_N, \mathfrak{U})$ is also congruence invariant. □

The geometric version of the Gibbs free energy function $G(\mathbf{X}; \mathfrak{E}_N, \mathfrak{U})$ as in (28), has the advantage that it is lower semi-continuous and if we construct the thermodynamic system $\mathfrak{S}_{\mathbf{X}}$ carefully, it is also a smooth function away from its infinity value set.

Theorem 6.2. *$G(\cdot\,; \mathfrak{E}_N, \mathfrak{U}) : \mathfrak{X}_{\mathfrak{U}} \to \mathbb{R}$ is lower semi-continuous.*

Proof. Let

$$D = \{\mathbf{X} = (\mathbf{x}_1, \cdots, \mathbf{x}_n) \in \mathbb{R}^3 \ : \ \mathbf{x}_i = \mathbf{x}_j, \ 1 \le i < j \le n\}. \tag{30}$$

On D, $G(\cdot\,; \mathfrak{E}_N, \mathfrak{U})$ takes value $+\infty$.

First, the Coulombs potential is continuous on $\mathfrak{X}_{\mathfrak{U}} \setminus D$, and lower semi-continuous on $\mathfrak{X}_{\mathfrak{U}}$. Second, by [42], the volume and area terms in $G(\mathbf{X}; \mathfrak{E}_N, \mathfrak{U})$ in (28) are even continuous. So $G(\mathbf{X}; \mathfrak{E}_N, \mathfrak{U})$ is lower semi-continuous. □

Theorem 6.3. *Let $G(\cdot\,; \mathfrak{E}_N, \mathfrak{U})(\mathfrak{X}_{\mathfrak{U}}) \subset \mathbb{R}$ be the range of $G(\cdot\,; \mathfrak{E}_N, \mathfrak{U}) : \mathfrak{X}_{\mathfrak{U}} \to \mathbb{R}$. There is a closed ball $B_M(\mathbf{X}_0)$ such that*

$$G(\cdot\,; \mathfrak{E}_N, \mathfrak{U})(\mathfrak{X}_{\mathfrak{U}}) = G(\cdot\,; \mathfrak{E}_N, \mathfrak{U})(B_M(\mathbf{X}_0) \cap \mathfrak{X}_{\mathfrak{U}}) \subset \mathbb{R}. \tag{31}$$

Proof. Because of bond lengths and bond angles must be respected as (1) and (2), and the molecular graph of $\mathfrak{U}$ is connected, after a congruence in $\mathbb{R}^3$ if necessary, any two conformations $\mathbf{X}$ and $\mathbf{Y}$ will satisfy $|\mathbf{X} - \mathbf{Y}| \le M$, where M is a positive constant. Let $B_M(\mathbf{X}_0) = \{\mathbf{X} \in \mathbb{R}^{3n} \ : \ |\mathbf{X} - \mathbf{X}_0| \le M\}$ be the closed ball in $\mathbb{R}^{3n}$, since $G(\mathbf{X}; \mathfrak{E}_N, \mathfrak{U}) = G(C(\mathbf{X}); \mathfrak{E}_N, \mathfrak{U})$, (31) is true. □

6.2 $G(\mathbf{X}; \mathfrak{C}_N, \mathfrak{U})$ Has At Least One Minimiser

A conformation $\mathbf{X}_N$ is a global minimiser if

$$G(\mathbf{X}_N; \mathfrak{C}_N, \mathfrak{U}) = \min_{\mathbf{X} \in \mathfrak{X}_{\mathfrak{U}}} G(\mathbf{X}; \mathfrak{C}_N, \mathfrak{U}).$$

A conformation $\mathbf{X}_L$ is a local minimiser means that for some $r > 0$,

$$G(\mathbf{X}_L; \mathfrak{C}_N, \mathfrak{U}) = \min_{\mathbf{X} \in \mathfrak{X}_{\mathfrak{U}} \cap B_r(\mathbf{X}_L)} G(\mathbf{X}; \mathfrak{C}_N, \mathfrak{U}).$$

Theorem 6.4. *The function $G(\mathbf{X}; \mathfrak{R}_N, \mathfrak{U})$ has at least one minimiser.*

Proof. The classical mathematics theorem is that, let $f : U \to \mathbb{R}$ be a lower semi-continuous function and $C \subset U$ be compact, then there is a $c \in C$, such that

$$f(c) = \min_{\mathbf{x} \in C} f(\mathbf{x}).$$

By Theorem 6.2, $G(\mathbf{X}; \mathfrak{C}_N, \mathfrak{U})$ is lower semi-continuous. Note that $\mathfrak{X}_U \cap B_M(\mathbf{X}) \subset \mathbb{R}^{3n}$ is compact for any $\mathbf{X} \in \mathbb{R}^3$. Thus by Theorem 6.3, there is a $\mathbf{X}_N \in \mathfrak{X}_{\mathfrak{U}} \cap B_M(\mathbf{X}_0)$ such that

$$G(\mathbf{X}_N; \mathfrak{C}_N, \mathfrak{U}) = \min_{\mathbf{X} \in \mathfrak{X}_{\mathfrak{U}} \cap B_M(\mathbf{X}_0)} G(\mathbf{X}; \mathfrak{C}_N, \mathfrak{U}) = \min_{\mathbf{X} \in \mathfrak{X}_{\mathfrak{U}}} G(\mathbf{X}; \mathfrak{C}_N, \mathfrak{U}),$$

the last equality is because (31). So $\mathbf{X}_N$ is a minimiser. □

There are two kinds of minimisers of $G(\mathbf{X}; \mathfrak{C}_N, \mathfrak{U})$, boundary minimisers and interior minimisers. If the minimiser $\mathbf{X}_B \in \partial \mathfrak{X}_{\mathfrak{U}}$, then it is a boundary minimiser. Note that, we can easily extend the construction of molecular surface to an open neighbourhood $U \subset \mathbb{R}^{3n}$ of $\mathbf{X}_B$, then $G(\mathbf{X}; \mathfrak{C}_N, \mathfrak{U})$ is a well defined function on U. If $G(\mathbf{X}; \mathfrak{C}_N, \mathfrak{U})$ is also smooth (has continuous first partial derivatives) then it is possible that $G(\mathbf{X}_B; \mathfrak{C}_N, \mathfrak{U}) > \inf_U G(\mathbf{X}; \mathfrak{C}_N, \mathfrak{U})$, because $\mathbf{X}_B$ is only a boundary minimiser.

6.3 Smoothness of $G(\mathbf{X}; \mathfrak{C}_N, \mathfrak{U})$

Theorem 6.5. *We can choose the atomic shape $B(\mathbf{x}_i)$ in subsection §§3.5, such that*
$G(\cdot\,; \mathfrak{C}_N, \mathfrak{U}) : \mathfrak{X}_{\mathfrak{U}} \backslash D \to \mathbb{R}$ smooth, i.e., has continuous first partial derivatives.

Proof. The Coulombs potential in $G(\mathbf{X}; \mathfrak{E}_N, \mathfrak{U})$ is smooth aways from D. Thus we only need prove that the volume and area terms in G as in (28) is smooth away from D.

In general, let Ω be a bounded domain in $\mathbb{R}^3$ and $\partial\Omega = \Sigma$ has a square integrable mean curvature vector $H(\mathbf{x})$, then the first variations of volumes and areas according to a vector field $L : \mathbb{R}^3 \to \mathbb{R}$ is (see, for example, [42])

$$\delta_L V(\Omega) = -\int_\Sigma L(\mathbf{x}) \bullet N(\mathbf{x}) \mathrm{d}\mathcal{H}^2(\mathbf{x}), \tag{32}$$

where $N(\mathbf{x})$ is the outward unit normal vector of Σ at $\mathbf{x} \in \Sigma$, $\bullet$ is the inner product in $\mathbb{R}^3$, and $\mathcal{H}^2$ is the Hausdorff measure of 2-dimensions.

The first variation of area is

$$\delta_L A(\Sigma) = -2\int_\Sigma [L(\mathbf{x}) \bullet H(\mathbf{x})] \mathrm{d}\mathcal{H}^2(\mathbf{x}). \tag{33}$$

The vector field L only needs to be square integrable on Σ to guarantee that the volume and area functions are smooth.

Thus to make $G(\mathbf{X}; \mathfrak{E}_N, \mathfrak{U})$ in (28) smooth, we further need: 1. $\Sigma_\mathbf{X}$ depends only on $\mathbf{X}$, i.e., if $\mathbf{X}(t) = (\mathbf{x}_1(t), \cdots, \mathbf{x}_n(t))$ is a family of conformation such that $\mathbf{X}(0) = \mathbf{X}$, there is a family of molecular surfaces $\Sigma(t) = \Sigma_{\mathbf{X}(t)}$, and the vector field defined on $\mathbf{X}$ by

$$L(\mathbf{x}_k) = \left.\frac{\mathrm{d}\mathbf{x}_k(t)}{\mathrm{d}t}\right|_{t=0}, \quad k = 1, \cdots, n, \tag{34}$$

has a unique extension to $\Sigma_\mathbf{X}$; 2. $\Sigma_\mathbf{X}$ has square integrable mean curvature.

When defining $\Omega_\mathbf{X}$ and $\Sigma_\mathbf{X} = \partial\Omega_\mathbf{X}$, to make the theory more general, we required only that there should be no room left inside $\Omega_\mathbf{X}$ for any molecules except $P_\mathbf{X} = \cup_{i=1}^n B(\mathbf{x}_i)$. That might make the obtained $\Sigma_\mathbf{X} = \partial\Omega_\mathbf{X}$ not satisfying the above two condition.

There are many surfaces satisfying above conditions. A particular one is the molecular surface, its two-dimensional project is showing in FIGURE 6.

Molecular surface $M_\mathbf{X}$ was defined in [35] and developed in [8]. It is generated by rolling a probe sphere of radius $d_w/2$ (radius of a water molecule) along $\partial P_\mathbf{X} = \partial(\cup_{i=1}^n B(\mathbf{x}_i, r_i))$. It is made up of pieces of faces that join at circular arcs. There are 3 kinds of faces: convex face (contained in some $\partial B(\mathbf{x}_i, r_i)$; concave face (a spherical triangle on the probe sphere when it is rolled into a position such that it touches simultaneously 3 balls); when the probe sphere touches simultaneously two balls the shorter great arc on the probe sphere between the two touching points sweep out a surface that is a saddle face. Convex faces are called the contact surface, the remanning is called the reentrant surface. These faces are defined by analytical expressions in terms of $\mathbf{X}$ and van der Waals radii r_is, and d_w. Thus any vector field can be uniquely extended to $M_{\mathbf{X}}$, as long as we know its definition on $\mathbf{X} = (\mathbf{x}_1, \cdots, \mathbf{x}_n)$. Each face is analytic, hence $M_{\mathbf{X}}$ is piecewise C^2 that will guarantee that $M_{\mathbf{X}}$ has square integrable mean curvature. In general, any closed surface (connected, bounded, and has no boundary, for example, a sphere) $S \subset \mathbb{R}^3$ will divide $\mathbb{R}^3$ into three parts,

$$\mathbb{R}^3 = \Omega_S \cup S \cup \Omega'_S, \quad \partial\Omega_S = \partial\Omega'_S = S, \tag{35}$$

where Ω_S is a bounded domain (connected open set) and Ω'_S a un-bounded domain, $\partial\Omega_S$ means the boundary of Ω_S.

If $M_{\mathbf{X}}$ is connected, then we can use $S = \Sigma_{\mathbf{X}} = M_{\mathbf{X}}$ in (35). If $M_{\mathbf{X}}$ has multiple connected components S_i, $1 \le i \le m$, such that S_1 is the largest component, i.e., all other components of $M_{\mathbf{X}}$ are contained in Ω_{S_1} (this is the case that $P_{\mathbf{X}}$ has $m-1$ cavities Ω_{S_i}, $i = 2, \cdots, m$, each is large enough to hold a water molecule, in this case $\partial P_{\mathbf{X}}$ has m components too), define $\Sigma_{\mathbf{X}} = M_{\mathbf{X}}$ and $\Omega_{\mathbf{X}} = \Omega_{S_1} \cap (\cup_{i=2}^m \Omega'_{S_i})$ and $\Omega'_{\mathbf{X}} = \Omega'_{S_1} \cup (\cup_{i=2}^m \Omega_{S_i})$. Thus, we always have

$$\mathbb{R}^3 = \Omega_{\mathbf{X}} \cup \Sigma_{\mathbf{X}} \cup \Omega'_{\mathbf{X}}, \quad \partial\Omega_{\mathbf{X}} = \partial\Omega'_{\mathbf{X}} = \Sigma_{\mathbf{X}}, \quad P_{\mathbf{X}} \subset \overline{\Omega}_{\mathbf{X}} = \Omega_{\mathbf{X}} \cup \Sigma_{\mathbf{X}}.$$

There is no room inside $\Omega_{\mathbf{X}} \setminus P_{\mathbf{X}}$ for water molecules, so we can take this $\Omega_{\mathbf{X}}$ to define $\mathfrak{S}_{\mathbf{X}}$ as before. Thus we can apply (32) and (33) to show that $G(\mathbf{X}; \mathfrak{E}_N, \mathfrak{U})$ has first partial derivatives. These partial derivatives are presented in integral form, which is continuous with $\Sigma_{\mathbf{X}}$ [42]. If we modify a little the definition of $M_{\mathbf{X}}$, i.e., allow the probe sphere rolling without blocked by non-local part of $P_{\mathbf{X}}$ to make the $M_{\mathbf{X}}$ being continuous with $\mathbf{X}$ by the generation of the molecular surface, so these first derivatives are continuous with $\mathbf{X}$. □

6.4 The Gradient $\nabla_{\mathbf{X}} G(\mathbf{X}; \mathfrak{E}_N, \mathfrak{U})$

Let L be a vector field defined on $\mathbf{X}$ by $\mathbf{X}_t$ as in (34). Let $\Sigma_{\mathbf{X}_t}$ be the molecular surface of $\mathbf{X}_t$. Define $f_{t,i} : \mathbb{R}^3 \to \mathbb{R}$ as

$$f_{t,i}(\mathbf{x}) = \mathrm{dist}(\mathbf{x}, \Sigma_{\mathbf{X}_t,i}) - \mathrm{dist}(\mathbf{x}, \Sigma_{\mathbf{X}_t} \setminus \Sigma_{\mathbf{X}_t,i}),$$

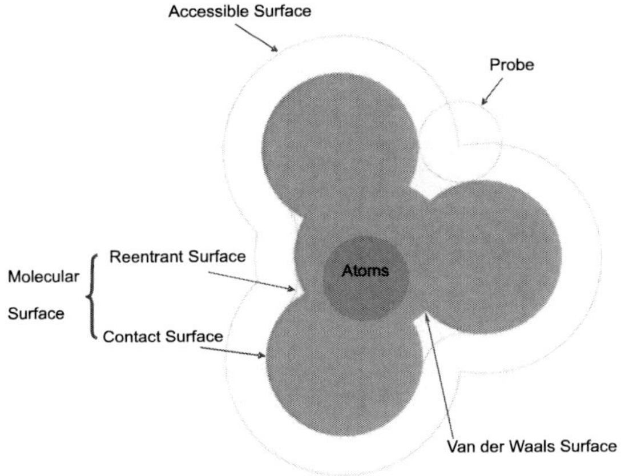

Figure 6. Two dimensional projection of molecular surface defined in [35] and solvent accessible surface defined in [23].

and let N be the outward unit normal of $\Sigma_{\mathbf{X}}$. On $\Sigma_{\mathbf{X}}$, define

$$\nabla_{\Sigma_{\mathbf{X}}} f_{0,i} = \nabla f_{0,i} - (\nabla f_{0,i} \bullet N)N, \quad f'_{0,i} = \left.\frac{\partial f_{t\,i}}{\partial t}\right|_{t=0}, \quad \frac{\mathrm{d} f_{0,i}}{\mathrm{d}t} = L \bullet \nabla f_{0,i} + f'_{0,i},$$

then let η be the unit outward conormal vector of $\partial\Sigma_{\mathbf{X},i}$ (normal to $\partial\Sigma_{\mathbf{X},i}$ but tangent to $\Sigma_{\mathbf{X}}$. Then, for any vector field L, we have

$$\begin{aligned} \delta_L A(\Sigma_{\mathbf{X},i}) &= -2\int_{\Sigma_{\mathbf{X}\,i}} L(\mathbf{x}) \bullet H(\mathbf{x}) \mathrm{d}\mathcal{H}^2(\mathbf{x}) \\ &\quad + \int_{\partial\Sigma_{\mathbf{X}\,i}} \left[L(\mathbf{x}) \bullet \eta(\mathbf{x}) - \frac{\frac{\mathrm{d} f_{0,i}}{\mathrm{d}t}}{|\nabla_{\Sigma_{\mathbf{X}}} f_{0,i}|} \right] \mathrm{d}\mathcal{H}^1(\mathbf{x}). \quad (36) \end{aligned}$$

$$\delta_L |\mathbf{x}_i - \mathbf{x}_j|^{-1} = \frac{(\mathbf{x}_j - \mathbf{x}_i) \bullet [(L(\mathbf{x}_i) - L(\mathbf{x}_j)]}{|\mathbf{x}_i - \mathbf{x}_j|^3}.$$

Remark 6.1. *Formula in (36) is a new mathematical result, its mathematical proof will appear elsewhere.*

Let $\mathbf{c} = (c_1, c_2, c_3) \in \mathbb{R}^3$, and $k \in \{1, 2, \cdots, n\}$,

$$\mathbf{x}_i(t) = \mathbf{x}_i + t\delta_{ik}\mathbf{c}, \quad L_{k\mathbf{c}}(\mathbf{x}_i) = \left.\frac{\mathrm{d}\mathbf{x}_i(t)}{\mathrm{d}t}\right|_{t=0} = \delta_{ik}\mathbf{c},$$

Then

$$\begin{aligned}
\mathbf{c}\bullet\nabla_{\mathbf{x}_k}G(\mathbf{X};\mathfrak{E}_N,\mathfrak{U}) &= \delta_{L_{k\mathbf{c}}}G(\mathbf{X};\mathfrak{E}_N,\mathfrak{U})\\
&= \omega_e\delta_{L_{k\mathbf{c}}}V(\Omega_{\mathbf{X}})+d_w\omega_e\delta_{L_{k\mathbf{c}}}A(\Sigma_{\mathbf{X}})+\sum_{i=1}^{k}\omega_i\delta_{L_{k\mathbf{x}}}A(\Sigma_{\mathbf{X},i}\\
&+ \delta_{L_{k\mathbf{c}}}\sum_{1\le i<j\le n}\frac{Z_iZ_j}{|\mathbf{x}_i-\mathbf{x}_j|}\\
&= -\omega_e\int_{\Sigma_{\mathbf{X}}}L_{k\mathbf{c}}(\mathbf{x})\bullet N(\mathbf{x})\mathrm{d}\mathcal{H}^2(\mathbf{x})-2d_w\omega_e\int_{\Sigma_{\mathbf{X}}}L_{k\mathbf{c}}(\mathbf{x})\bullet H(\mathbf{x})\mathrm{d}\mathcal{H}^2(\mathbf{x})\\
&\quad -2\sum_{i=1}^{k}\omega_i\int_{\Sigma_{\mathbf{X}}}L_{k\mathbf{c}}(\mathbf{x})\bullet H(\mathbf{x})\mathrm{d}\mathcal{H}^2(\mathbf{x})\\
&\quad +\sum_{i=1}^{k}\int_{\partial\Sigma_{\mathbf{X}\,i}}\left[L_{k\mathbf{c}}(\mathbf{x})\bullet\eta(\mathbf{x})-\frac{\frac{\mathrm{d}f_{0,i}}{\mathrm{d}t}}{|\nabla_{\Sigma_{\mathbf{X}}}f_{0,i}|}\right]\mathrm{d}\mathcal{H}^1(\mathbf{x})\\
&\quad -\sum_{i\neq k}\frac{Z_kZ_i(\mathbf{x}_k-\mathbf{x}_i)\bullet\mathbf{c}}{|\mathbf{x}_k-\mathbf{x}_i|^3}.
\end{aligned}$$

Let $\mathbf{x}_k = (x_k^1, x_k^2, x_k^3)$, $k = 1,\cdots,n$, $\mathbf{e}_1 = (1,0,0)$, $\mathbf{e}_2 = (0,1,0)$, $\mathbf{e}_3 = (0,0,1)$, then

$$\frac{\partial G(\mathbf{X};\mathfrak{E}_N,\mathfrak{U})}{\partial x_k^j} = \mathbf{e}_j\bullet\delta_{L_{k\mathbf{e}_j}}G(\mathbf{X};\mathfrak{E}_N,\mathfrak{U}),\quad j=1,2,3.$$

The kth gradient is

$$\nabla_{\mathbf{x}_k}G(\mathbf{X};\mathfrak{E}_N,\mathfrak{U}) = \left(\frac{\partial G(\mathbf{X};\mathfrak{E}_N,\mathfrak{U})}{\partial x_k^1},\frac{\partial G(\mathbf{X};\mathfrak{E}_N,\mathfrak{U})}{\partial x_k^2},\frac{\partial G(\mathbf{X};\mathfrak{E}_N,\mathfrak{U})}{\partial x_k^3}\right).$$

The total gradient is

$$\nabla_{\mathbf{X}}G(\mathbf{X};\mathfrak{E}_N,\mathfrak{U}) = (\nabla_{\mathbf{x}_1}G(\mathbf{X};\mathfrak{E}_N,\mathfrak{U}),\cdots,\nabla_{\mathbf{x}_n}G(\mathbf{X};\mathfrak{E}_N,\mathfrak{U})).$$

6.5 The Gradient $\nabla_\Phi G_{[\mathbf{X}]}G(\Phi;\mathfrak{E}_N,\mathfrak{U})$

Let $\mathbf{b}_k = \mathbf{b}_{i_kj_k} = \frac{\mathbf{x}_{j_k}-\mathbf{x}_{i_k}}{|\mathbf{x}_{j_k}-\mathbf{x}_{i_k}|}$, $k=1,\cdots,L$, be the axis of the bond b_k. Take the derivative against θ in (7), we have

$$R_k'(X) = R_{i_kj_k\theta}'(X)|_{\theta=0} = (X-\mathbf{x}_{i_k})\wedge\mathbf{b}_k.$$

Let $G_{b_k,1} \cup G_{b_k,2} = G_{b_k}$, define the vector field L_k on $\mathbf{X}$ by

$$L_k(\mathbf{x}_i) = \begin{cases} (\mathbf{x}_i - \mathbf{x}_{i_k})\wedge\mathbf{b}_k, & \text{if } \mathbf{a}_i \in V(G_{b_k,2}); \\ 0, & \text{if } \mathbf{a}_i \in V(G_{b_k,1}). \end{cases}$$

Then

$$\delta_{L_k}|\mathbf{x}_i - \mathbf{x}_j|^{-1} = -\frac{(\mathbf{x}_i - \mathbf{x}_j)\bullet[L_k(\mathbf{x}_i) - L_k(\mathbf{x}_j)]}{|\mathbf{x}_i - \mathbf{x}_j|^3}.$$

If $\mathbf{a}_i$ and $\mathbf{a}_j$ both belong to $V(G_{b_k,1})$, then $L(\mathbf{x}_i) = L(\mathbf{x}_j) = \mathbf{0}$, $\delta_{L_k}|\mathbf{x}_i - \mathbf{x}_j|^{-1} = 0$. If $\mathbf{a}_i$ and $\mathbf{a}_j$ both belong to $V(G_{b_k,2})$, then $L(\mathbf{x}_i) - L(\mathbf{x}_j) = (\mathbf{x}_i - \mathbf{x}_j)\wedge\mathbf{b}_k$, hence $(\mathbf{x}_i - \mathbf{x}_j)\bullet[L_k(\mathbf{x}_i) - L_k(\mathbf{x}_j)] = 0$, also $\delta_{L_k}|\mathbf{x}_i - \mathbf{x}_j|^{-1} = 0$. This is consistent with the intuition and is also a proof that transformer rotation does not change bond lengths and angles.

The gradient of $G_{[\mathbf{X}]}$ then is

$$\begin{aligned} \nabla_\Phi G_{[\mathbf{X}]}(\Phi; \mathfrak{E}_N, \mathfrak{U}) &= (\delta_{L_1} G_{[\mathbf{X}]}(\Phi; \mathfrak{E}_N, \mathfrak{U}), \cdots, \delta_{L_L} G_{[\mathbf{X}]}(\Phi; \mathfrak{E}_N, \mathfrak{U})) \\ &= \left(\frac{\partial G_{[\mathbf{X}]}(\Phi; \mathfrak{E}_N, \mathfrak{U})}{\partial\phi_1}, \cdots, \frac{\partial G_{[\mathbf{X}]}(\Phi; \mathfrak{E}_N, \mathfrak{U})}{\partial\phi_L}\right) \end{aligned} \quad (37)$$

Remark 6.2. *Let* $\prod_{i=1}^{L}[0, 2\pi] = \overbrace{[0, 2\pi] \times \cdots \times [0, 2\pi]}^{L}$, *Since* $[\mathbf{X}]$ *is compact and* $G_{[\mathbf{X}]}(\cdot\,; \mathfrak{E}_N, \mathfrak{U}) : \prod_{i=1}^{L}[0, 2\pi] \to \mathbb{R}$ *is lower semi-continuous, there is at one least minimiser* $\Phi_0 \in \prod_{i=1}^{L}[0, 2\pi]$. *Since* $\prod_{i=1}^{L}[0, 2\pi]$ *has no boundary points, it must be that*

$$\nabla_\Phi G_{[\mathbf{X}]}(\Phi_0; \mathfrak{E}_N, \mathfrak{U}) = \mathbf{0} = \overbrace{(0, \cdots, 0)}^{L}. \quad (38)$$

Moreover, let $\mathbf{X}_0 = \mathbf{Y}(\Phi_0)$, *if* $\mathbf{X} \in \mathfrak{X}_{\mathfrak{U}}^{\circ}$, *then* $[\mathbf{X}] \subset \mathfrak{X}_{\mathfrak{U}}^{\circ}$. *Hence even* $\mathbf{X}_0$ *is not a minimiser of* $G(\mathbf{X}; \mathfrak{E}_N, \mathfrak{U})$, $\mathbf{X}_0$ *will still be a stationary point of* $G_{[\mathbf{X}]}(\Phi; \mathfrak{E}_N, \mathfrak{U})$.

6.6 Determine The Coefficients

Smoothness has an advantage, it can help us find the values of the coefficients ω_e and ω_i's in (28).

Actually, ω_e and ω_i's can be determined by applying the thermodynamic principle and solving linear equations.

Taking from data base a known native structure $\mathbf{X}_N$ of a globular protein $\mathfrak{U}$. By the Thermodynamic Principle, $\mathbf{X}_N$ is a stabiliser of $G(\mathbf{X}; \mathfrak{E}_N, \mathfrak{U})$ since it has minimal Gibbs free energy. So we have $\nabla_\Phi G_{[\mathbf{X}_N]}(\Phi(\mathbf{X}_N); \mathfrak{E}_N, \mathfrak{U}) = \mathbf{0}$.

It will always be $k < L$, since there are far more rotatable covalent bonds than hydrophobicity classes in any protein. Select $k+1$ rotatable bonds of $\mathfrak{U}$, say they are $b_1, \cdots, b_{k+1}$. Then

$$\frac{\partial G_{[\mathbf{X}_N]} G(\Phi(\mathbf{X}_N); \mathfrak{E}_N, \mathfrak{U})}{\partial \phi_i} = 0, \quad i = 1, \cdots, k+1$$

implies that

$$\omega_e \delta_{L_i} V(\Omega_{\mathbf{X}_N}) + d_w \omega_e \delta_{L_i} A(\Sigma_{\mathbf{X}_N}) + \sum_{j=1}^{k} \omega_j \delta_{L_i} A(\Sigma_{\mathbf{X}_{N,j}}) + \sum_{1 \le k < l} \delta_{L_i} \frac{Z_k Z_l}{|\mathbf{x}_{N,k} - \mathbf{x}_{N,l}|} = 0. \quad i = 1, \cdots, k+1.$$

There are $k+1$ equations, so in general we can solve the $k+1$ coefficients ω_e, ω_i's, by selecting $k+1$ rotatable bonds if necessary.

Note that even $\mathbf{X}_N$ is only a boundary minimiser, the partial derivations about the dihedral angles make sense, since $G_{[\mathbf{X}_N]}(\cdot\,; \mathfrak{E}_N, \mathfrak{U}) \prod_{i=1}^{L} [0, 2\pi] \to \mathbb{R}$ is away from the set D defined in (31).

6.7 The Graph of $G(\mathbf{X}; \mathfrak{E}_N, \mathfrak{U})$

Let

$$g_{\mathfrak{E}_N, \mathfrak{U}}(G) = \{(\mathbf{X}, z) \in \mathfrak{X}_{\mathfrak{U}} \times \mathbb{R} \,:\, z = G(\mathbf{X}; \mathfrak{E}_N, \mathfrak{U})\} \subset \mathfrak{X}_{\mathfrak{U}} \times \mathbb{R} \tag{39}$$

be the graph of $G(\cdot\,; \mathfrak{E}_N, \mathfrak{U}) : \mathfrak{X}_{\mathfrak{U}} \to \mathbb{R}$. It is a $3n$ dimensional hyper-surface in the $(3n+1)$ dimensional space $\mathfrak{X}_{\mathfrak{U}} \times \mathbb{R}$.

The graph of $g_{\mathfrak{E}_N, \mathfrak{U}}(G)$ is called the free energy landscape in protein folding community, but nobody has emphasised the definition (39), because without knowing the function G, it is useless. Instead, people use energy landscape language rather want to get the Gibbs free energy value of a conformation or a class of conformations, see [39].

Any molecule $\mathfrak{U}$, not necessarily a protein, has a Gibbs free energy function in terms of its environment. So we will use $g_{\mathfrak{E}, \mathfrak{U}}(G)$ to denote such a graph of a Gibbs free energy function G of a molecule $\mathfrak{U}$ in an environment $\mathfrak{E}$.

If G has a unique minimiser, or just a few grouped in nearby positions, it is said that $g_{\mathfrak{C},\mathfrak{U}}(G)$ looks funnel like, though in reality no body can visualise such a high dimensional hyper-surface. Otherwise, the $g_{\mathfrak{C},\mathfrak{U}}(G)$ is rugged, meaning that there are many minimisers, boundary or interior, global or local. Given the huge dimensions of conformation space of $\mathfrak{U}$, the rugged graph $g_{\mathfrak{C}_N,\mathfrak{U}}$ that leads to no particular conformation is preferred in nature, these happen for almost all heteropolymers (for example, a reshuffled peptide chain of a protein), except of proteins [39]. We will consider them as general heteropolymers. The ruggedness of $g_{\mathfrak{C}_N,\mathfrak{U}}(G)$, where $\mathfrak{U}$ is a general heteropolymer, can be explained by the concept of folding paths, see §§7.2.

7. Explanations

A scientific theory should be able to explain natural phenomena that the theory was supposed to address. Using the derived smooth Gibbs free energy function $G(\mathbf{X}; \mathfrak{C}_N, \mathfrak{U})$ we can explain many natural phenomena associated with protein folding. We can see that in any of the following cases, it does not violate any known physical laws or contradict to known facts. And in some cases, the single molecule theory explained better than old ones, such as the bottom-up explanation of protein structures and chameleon sequences.

7.1 A Holistic View of Protein Structures and Protein Folding

Classical view of protein structure is that there are four levels of structure, primary structure, the amino acid sequence; secondary structure, such as the α helices and β strands and sheets, β turns, all have regular hydrogen bond patterns; tertiary structure, assembling secondary structures to form a compact structure, two secondary structures are connected by a loop that does not have regular hydrogen bond pattern; and finally the quaternary structure, only for multi-peptide chain proteins, tightly packing all tertiary structures peptide chains. See for example, [7].

This view is a bottom-up view, i.e., primary structure determines secondary structures; assembly the secondary structures to form a compact tertiary structure; and the quaternary structure are the assembly of tertiary structures. It is rather like building a house by assembly method, all walls, stairs, floors, roofs, are separately manufactured previously, then assembled together to build a predesigned house. Therefore, an inference of this bottom-up view is that short-range conformational energies determining secondary propensities of short peptide chains, see [4].

A holistic view of protein structure coming from the Gibbs free energy function in (28). All concerned are volume, area, hydrophobicity class surface areas, and the Coulombs potential. Except the atomic centre distance, all of them are global geometric quantities of the conformation $P_{\mathbf{X}} \subset \mathbb{R}^3$ and space distributions of hydrophobicity classes. Local geometries, except the atomic centres' distance, never appear in the Gibbs free energy formula. Thus if in the native structures of proteins there are regular secondary structures, they must be the result of minimisation of Gibbs free energy among all conformations of the protein molecule, except to the Coulombs potential that prohibits two atomic centres too close to each other, no short-range conformational energies were involved. Thus the holistic structure view is a top-down view, to be minimal in the Gibbs free energy forces various parts of the peptide chain take some regular geometric shape that also fit the hydrophobicity distribution pattern, such as secondary structures. In physiological environment, depending on the whole amino acid sequences of proteins, a short peptide chain in native structures of different proteins may appear as different secondary structures, or no secondary structures at all, just a loop.

In this holistic view, a protein may have a native structure but does not have secondary structures, such as the intrinsically disordered proteins, [11]. In fact, as long as we can infer that an ensemble of protein molecules all have the same specific structure in physiological environment, we say this protein has a native structure, it does not matter whether or not there are secondary structures in that specific structure.

Which view can explain more phenomena in protein folding without self-contradiction? The chameleon sequences can be used as a test.

Chameleon sequences are short peptide chains up to 8 residues, in different proteins, the sequence may take different secondary structures, α helix becomes β strand, and *vice versa*, [17]. Holistic view explains them very well, the whole amino acid sequence of a protein and environment decide the Gibbs free energy and hence the minimisers' (native structures') secondary structures of various subsequence of the whole sequence. Therefore, it is quite possible that the same short peptide chain takes different secondary structures in native structures of different proteins.

The propensity theory, whether depends on short-range conformational energies of other theory, can only give frequency statistics of these chameleon sequences appear as various of secondary structures, but can never explain why they change, anyhow, the short-range is the same.

On the other hand, if there really are short-range conformational energies that determine secondary propensities of short peptide chain, there will be no chameleon sequences. The bottom-up view can not explain chameleon sequences at all. The best it can do is counting the frequencies of various secondary structures a chameleon sequence takes.

Further evidence of the top down view has grabbed the essential of protein structure is a minimisation of hydrophobic surface area simulation. In [15], calculations of shrinking the hydrophobic surface area $A(\Sigma_{\mathbf{X}h})$ was performed. The result is that hydrogen bonds and regular secondary structures such as α helices, β strands, and β turns appeared with statistic significance, without calculating and pursuing any local geometry such as atomic distance and dihedral angles. One explanation of this result is that secondary structures and hydrogen bonds actually are the result of hiding hydrophobic classes to form a core of the native structure, or in other words, the result of reduced Gibbs free energy; not the other way, i.e., arranging these secondary structures and hydrogen bonds to pursue a compact tertiary structure. Looking from the view of point of folding force, the explanation is just that regular secondary structures and hydrogen bond patterns will reduce the Gibbs free energy of the whole.

On the other hand, bottom up structure view implies that energies can be calculated locally regardless the whole, but it has been shown that pairwise energy simulation without specify hydrogen bonding cannot produce secondary structures, [19].

Gibbs free energy function in (28) forces another holistic view on protein folding, i.e., it is the global geometry of the conformation in $\mathbb{R}^3$, the $P_{\mathbf{X}}$, the $\Omega_{\mathbf{X}}$, that count in evaluating the Gibbs free energy. If we take a coarse-grained model of conformation $P_{\mathbf{X}}$, i.e., various methods that removes many atoms in $\mathfrak{U}$, some put side chains as a ball of different size, some even reduce a residue as a ball, cannot be used as model. If using a coarse-grained model, the Gibbs free energy will be quite different from that obtained from a full atom model. Then, minimisers will become non-minimisers and *vice versa*. Therefore, protein structure prediction by looking for minimisers of the Gibbs free energy function cannot use coarse-grained model of protein conformation. But many believe that use coarse-grained model to predict proteins' native structures not only possible, but also has advantages, see for example, [39]. Looking in detail of [39], the advantages of coarse-grained model come from reducing complexity of calculations, but never discussed the accuracy.

7.2 Folding Force and Folding Path

The so called Levinthal's paradox is actually a proof by contradiction in Levinthal's lecture [25]. In which he argued that if a protein folds by randomly searching all possible conformations (he estimated that for a protein of 2,000 atoms the number of conformations would be 10^{300}, because he only counted discrete dihedral angle differences and kept the bond lengths and angles invariant). In natural, proteins fold in about several seconds time frame, even with the fastest random searching, in several seconds' time the protein can only search at most about 10^8 conformations. Thus, there should request at least in the order of 10^{280} years time to search all conformations. Apparently that is impossible, the whole universe's life is only in the order 10^{150} years. By this contradiction, Levinthal concluded that there must be folding force pushing the protein fold along a folding path. Indeed, one year early, in [24] he suggested that there exist folding paths such that protein will fold only along a tiny part of the whole conformation space. That tiny part is the folding path. The folding path concept completely ruled out the random searching folding. The remaining is to find out what is the folding force and what is the path.

Remark 7.1. *If count smooth changes of conformations, then there will be infinity many conformations. One will immediately realise that it is nonsense to estimate random searching time at all, because that one can never finish an infinite job in a finite time. Therefore, random searching will be dismissed at the first place.*

Due to paucity of knowledge and theory at the early stage of the protein folding study, Levinthal could not pinpoint what is the folding force and what is the folding path. But he was definitely on the correct track. Unfortunately, many ones really fell into the random searching trap. Many argued that with a little deliberate choice in each random step, the protein will quickly find its native structure. This is rather anthropomorphise proteins, as if they had thought and known what their destinations are, and could subjectively make their choices in route to their destinations.

What Levinthal thought is in the first principles of physics. Matters move because of there are forces pushing them according fundamental physical laws. Therefore, there is a folding force, and from any conformation $\mathbf{X}_0$, there is a unique folding path $\mathbf{X}(t)$, such that $\mathbf{X}(0) = \mathbf{X}_0$. The problem is to find them, and make explanation without contradict to any fundamental physical laws and known facts. Knowing the gradient of Gibbs free energy, we can answer these questions.

Actually, the folding force is $-\nabla_{\mathbf{X}} G(\mathbf{X}; \mathfrak{E}_N, \mathfrak{U})$ and $-\nabla_{\Phi} G_{[\mathbf{X}]}(\Phi; \mathfrak{E}_N, \mathfrak{U})$. For instance, folding force acted on an atom $\mathbf{a}_i$ in the conformation $\mathbf{X}$ is given by $\mathbf{F}_i = -\nabla_{\mathbf{x}_i} G(\mathbf{X}; \mathfrak{E}_N, \mathfrak{U})$. Similarly, the force that causes a transformer rotation around a rotatable bond b_i is given by $\mathbf{F}_i = -\delta_{L_i} G_{[\mathbf{X}]} G(\Phi; \mathfrak{E}_N, \mathfrak{U})$.

Knowing the folding force, we can give the folding paths. Let $= \prod_{i=1}^{L} [0, 2\pi]$ $= \overbrace{[0, 2\pi) \times \cdots \times [0, 2\pi]}^{L}$. Given any initial conformation $\mathbf{X}_0 \in \mathfrak{X}^{\circ}_{\mathfrak{U}}$, $\Phi_0 = \Phi(\mathbf{X}_0) \in \prod_{i=1}^{L} [0, 2\pi]$, the folding path is $\mathbf{X}(t, \mathbf{X}_0) \in \mathfrak{X}_{\mathfrak{U}}$ such that $\mathbf{X}(0, \mathbf{X}_0) = \mathbf{X}_0$, and $\Phi(t, \Phi_0)$ such that $\Phi(0, \Phi_0) = \Phi_0$,

$$\frac{\mathrm{d}\mathbf{X}(t, \mathbf{X}_0)}{\mathrm{d}t} = -\nabla_{\mathbf{X}} G(\mathbf{X}(t, \mathbf{X}_0); \mathfrak{E}_N, \mathfrak{U}), \tag{40}$$

$$\frac{\mathrm{d}\Phi(t, \Phi_0)}{\mathrm{d}t} = -\nabla_{\Phi} G_{[\mathbf{X}_0]}(\Phi(t, \Phi_0); \mathfrak{E}_N, \mathfrak{U}). \tag{41}$$

We will call $\mathbf{X}(t, \mathbf{X}_0)$ the full folding path and $\Phi(t, \Phi_0)$ the rotation folding path. The notations $\mathbf{X}(t, \mathbf{X}_0)$ and $\Phi(t, \Phi_0)$ emphasise that the folding path depend on the initial conformation. Let $G(t) = G(\mathbf{X}(t, \mathbf{X}_0); \mathfrak{E}_N, \mathfrak{U})$ be the Gibbs free energy along the full folding path, then

$$\begin{aligned} G'(t) &= \frac{\mathrm{d}G(t)}{\mathrm{d}t} = \nabla_{\mathbf{X}} G(\mathbf{X}(t, \mathbf{X}_0); \mathfrak{E}_N, \mathfrak{U}) \bullet \frac{\mathrm{d}\mathbf{X}(t, \mathbf{X}_0)}{\mathrm{d}t} \\ &= -|\nabla_{\mathbf{X}} G(\mathbf{X}(t); \mathfrak{E}_N, \mathfrak{U})|^2 \leq 0. \end{aligned}$$

Let $G_{[\mathbf{X}_0]}(t) = G_{[\mathbf{X}_0]}(\Phi(t, \Phi_0); \mathfrak{E}_N, \mathfrak{U})$ be the Gibbs free energy along the rotation folding path, then

$$\begin{aligned} G'_{[\mathbf{X}_0]}(t) &= \frac{\mathrm{d}G_{[\mathbf{X}_0]}(t)}{\mathrm{d}t} = \nabla_{\Phi} G_{[\mathbf{X}_0]}(\Phi(t, \Phi_0); \mathfrak{E}_N, \mathfrak{U}) \bullet \frac{\mathrm{d}\Phi(t, \Phi_0)}{\mathrm{d}t} \\ &= -|\nabla_{\Phi} G_{[\mathbf{X}_0]}(\Phi(t, \Phi_0); \mathfrak{E}_N, \mathfrak{U})|^2 \leq 0. \end{aligned}$$

Since $G(\mathbf{X}; \mathfrak{E}_N, \mathfrak{U})$ is bounded below, the folding path $\mathbf{X}(t, \mathbf{X}_0)$ will stop at a minimiser $\mathbf{X}(t_0, \mathbf{X}_0)$. If it is a interior minimiser, then $G'(t_0) = 0$. Otherwise $\mathbf{X}(t_0) \in \partial \mathfrak{X}_{\mathfrak{U}}$ is a boundary minimiser. Hence, starting with any conformation

$\mathbf{X}_0$, the full folding path $\mathbf{X}(t, \mathbf{X}_0)$ will lead us to a minimiser $\mathbf{X}(t_0)$, and then stay there. The advantage of the rotation folding path is that starting at a interior conformation $\mathbf{X} \in \mathfrak{X}^\circ_{\mathfrak{U}}$, we will reach a minimiser Φ_M, $\mathbf{Y}(\Phi_M) \in [\mathbf{X}] \subset \mathfrak{X}^\circ_{\mathfrak{U}}$, and always have $\nabla_\Phi G_{[\mathbf{X}]}(\Phi_M; \mathfrak{E}_N, \mathfrak{U}) = \mathbf{0}$.

7.2.1 Why Heteropolymers, Except Proteins, Do Not Have Native Structures?

It is known that heteropolymers do not have native structure, i.e., they generally have many conformations in equilibrium in an environment. Ruggedness of free energy landscape and the folding path above can explain why.

Every heteropolymer $\mathfrak{V}$ has its own Gibbs free energy function $G(\mathbf{X}; \mathfrak{E}, \mathfrak{V})$ in an environment $\mathfrak{E}$. Therefore, each conformation $\mathbf{X}$ of $\mathfrak{V}$ in the environment $\mathfrak{E}$ should have a folding path $\mathbf{X}(t, \mathbf{X})$ such that $\mathbf{X}_M = \mathbf{X}(t_0, \mathbf{X})$, for some $t_0 > 0$, is a minimiser of G.

For each minimiser $\mathbf{X}_M$ of $G(\mathbf{X}; \mathfrak{E}, \mathfrak{V})$, boundary or interior, there is a maximum converging domain $O_{\mathbf{X}_M} \subset \mathfrak{X}_{\mathfrak{U}}$ such that if $\mathbf{X}_0 \in O_{\mathbf{X}_M}$, then $\mathbf{X}(t_0, \mathbf{X}_0) = \mathbf{X}_M$ for some $t_0 > 0$. But if $\mathbf{X}_0 \notin \overline{O}_{\mathbf{X}_M}$, $\mathbf{X}(t, \mathbf{X}_0)$ will lead to another minimiser rather than $\mathbf{X}_M$.

Let S be an ensemble of identical molecules of a heteropolymer $\mathfrak{V}$. If all initial conformations in S are concentrated in one maximum converging domain $O_{\mathbf{X}_N}$, then $\mathfrak{V}$ should have a native structure $\mathbf{X}_N$, even the graph $g_{\mathfrak{E},\mathfrak{V}}(G)$ is rugged. Thus S does not have a native structure means that: the graph $g_{\mathfrak{E},\mathfrak{V}}(G)$ is rugged; and initial conformations in S split widely, i.e., for each minimiser $\mathbf{X}_S$ of $G(\mathbf{X}; \mathfrak{E}, \mathfrak{V})$, there are substantial quantities of initial conformations contained in $O_{\mathbf{X}_S}$, therefore, $\mathbf{X}(t_{\mathbf{X}_0}, \mathbf{X}_0) = \mathbf{X}_S$ for all $\mathbf{X}_0 \in O_{\mathbf{X}_S}$. Thus, at the end, substantial number of molecules in S takes each of all possible minimiser conformations and stay there. Since $g_{\mathfrak{E},\mathfrak{V}}(G)$ is rugged, in environment $\mathfrak{E}$, $\mathfrak{V}$ does not have a native structure.

7.2.2 Why Proteins Usually Have Native Structures

Proteins are really special, in physiological environment, and even physiological environment with a little variation, proteins usually have specific structures, their native structures. The arguments given in last subsection show that the reason can only be the two alternatives.

One, the graph $g_{\mathfrak{E}_N,\mathfrak{U}}(G)$ is funnel like, i.e., has only one minimiser or just several minimisers close to each other in space. This seems is the current explanation.

But via the given folding path, there is another explanation which seems nobody has argued before. That is, the initial conformations of a protein $\mathfrak{U}$ are concentrated in a specific maximum converging domain $O_{\mathbf{X}_M}$, so in folding all finally fall to $\mathbf{X}_M$, the native structure is $\mathbf{X}_M$, even though it may be just a local minimiser and $g_{\mathfrak{C}_N,\mathfrak{V}}(G)$ is rugged.

The necessary condition of the second alternative is that newly synthesised proteins have similar conformations such that they all belong to the same maximum converging domain. We need highly sophisticated single molecule test method to test this alternative.

Because of the high dimensions and the complexity of $G(\mathbf{X};\mathfrak{C}_N,\mathfrak{U})$, it is impossible to visualise the graph $g_{\mathfrak{C}_N,\mathfrak{U}}(G)$, the much cited funnel picture had actually been pointed as wrong in fundamental concept, i.e., the picture implies that the smaller the entropy, the smaller the Gibbs free energy, [20]. We have to cite other known facts to infer which alternatives is the case for a specific protein.

The globular protein ribonuclease has 105 different disulphide bond patterns, and the native structure belongs one of them. By adding denature agents, the protein lost roughly 99% of its functional capacity. So one infers that in the denatured environment, the protein takes every one of the 105 disulphide bond patterns, only about 1% are in the native structure disulphide bond pattern. But the protein can refold to its native structure spontaneously once one removes the denature agents [1]. This fact infers that the graph $g_{\mathfrak{C}_R,\mathfrak{V}}G$, where $\mathfrak{C}_R$ is the refolding environment for ribonuclease, must be funnel like, because all initial conformations split to all possibilities evenly but still they all fall in just one maximum converging domain.

We still cannot rule out that for some proteins, the concentration on native structure is due to the second alternative. The final answer maybe is that both alternatives are true, some proteins have funnel like graph $g_{\mathfrak{C}_N,\mathfrak{U}}(G)$, some have similar initial conformations.

Why proteins are so special among heteropolymers? The answer is that in evolution, protein molecules have to have specific structures, therefore, the amino acid sequences have to be specified to make various moieties with different hydrophobicity levels are arranged in 3-dimensional space to form the specific, native structure. How specified of the amino acid sequences of proteins? It was shown that randomly picking up an amino acid sequence, its probability of being a protein's amino acid sequence is essentially zero. For example, it is estimated in [16] that for random peptide chain $\mathbf{A}$ of 400 residues, the probability $p(\mathbf{A})$ of $\mathbf{A}$ is a protein's peptide chain is

$$P(\mathbf{A}) \leq 10^{60}/10^{520} = 10^{-460},$$

it is practically zero.

With a chance of less than 10^{-460}, natural selection is very choosy indeed. But this does not mean that natural selection is among all the possible amino acid sequences, randomly try each one, in that sense there is no chance of life evolution. On the contrary, there must be certain physic-chemical law that put a strict restriction on the amino acid sequences that can naturally occur, i.e., only a very very tiny part of amino acid sequences have appeared in nature and the natural selection of evolution of life is among them instead of all possible amino acid sequences.

7.3 Importance of Environments

Let $\mathfrak{U}_1$ and $\mathfrak{U}_2$ be two protein molecules, and they stay in their native structures $\mathbf{X}_1$ and $\mathbf{Y}_1$. The 3-dimensional conformations $P_{\mathbf{X}_1}$ and $P_{\mathbf{Y}_1}$ are contained in their tailor made thermodynamic systems $\mathfrak{S}_{\mathbf{X}_1}$ and $\mathfrak{S}_{\mathbf{Y}_1}$ respectively in the same cell and the same environment $\mathfrak{E}_N$. Besides $P_{\mathbf{X}_1}$ and $P_{\mathbf{Y}_1}$, we will call $\mathcal{R}_{\mathbf{X}_1}$ and $\mathcal{R}_{\mathbf{Y}_1}$ their private domains respectively.

In the beginning, $\mathfrak{S}_{\mathbf{X}_1} \cap \mathfrak{S}_{\mathbf{Y}_1} = \emptyset$. Now suppose that there is a congruence $C : \mathbb{R}^3 \to \mathbb{R}^3$ brings $\mathfrak{S}_{\mathbf{Y}_1}$ to $\mathfrak{S}_{\mathbf{X}_1}$, i.e., $\mathfrak{S}_{\mathbf{X}_1} \cap C(\mathfrak{S}_{\mathbf{Y}_1}) \neq \emptyset$. For simplicity, we drop the congruence C, then $\mathfrak{S}_{\mathbf{X}_1} \cap \mathfrak{S}_{\mathbf{Y}_1} \neq \emptyset$. Since it is always $P_{\mathbf{X}_1} \cap P_{\mathbf{Y}_1} = \emptyset$, it must be that

$$\mathcal{R}_{\mathbf{X}_1} \cap \mathcal{R}_{\mathbf{Y}_1} \neq \emptyset. \tag{42}$$

In this case, we say that the private domains of $\mathbf{X}_1$ and $\mathbf{Y}_1$ are invaded.

Consider $P_{\mathbf{X}_1} \cup P_{\mathbf{Y}_1}$ of the conformation $(\mathbf{X}_1, \mathbf{Y}_1) \in \mathfrak{X}_{\mathfrak{U}_1} \times \mathfrak{X}_{\mathfrak{U}_2} \supset \mathfrak{X}_{(\mathfrak{U}_1, \mathfrak{U}_2)}$, i.e., we see $(\mathfrak{U}_1, \mathfrak{U}_2)$ as a combined entity. Then there should be a new thermodynamic system tailor made for the conformation $(\mathbf{X}_1, \mathbf{Y}_1) \in \mathfrak{X}_{(\mathfrak{U}_1, \mathfrak{U}_2)}$. Then we have a new Gibbs free energy function

$$G(\cdot\,; \mathfrak{E}_N, (\mathfrak{U}_1, \mathfrak{U}_2)) : \mathfrak{X}_{(\mathfrak{U}_1, \mathfrak{U}_2)} \to \mathbb{R}, \tag{43}$$

to which the conformation $(\mathbf{X}_1, \mathbf{Y}_1)$ is not a minimiser. The two molecules will fold together along a folding path $(\mathbf{X}, \mathbf{Y})(t, (\mathbf{X}_1, \mathbf{Y}_1))$ until they reach a minimiser $(\mathbf{X}_2, \mathbf{Y}_2)$. Therefore, the two native structures $\mathbf{X}_1$ and $\mathbf{Y}_1$ has been changed to non-minimisers $\mathbf{X}_2$ and $\mathbf{Y}_2$ respectively as shown in FIGURE 7. But $(\mathbf{X}_2, \mathbf{Y}_2)$ is a minimiser of the new Gibbs free energy function in (43).

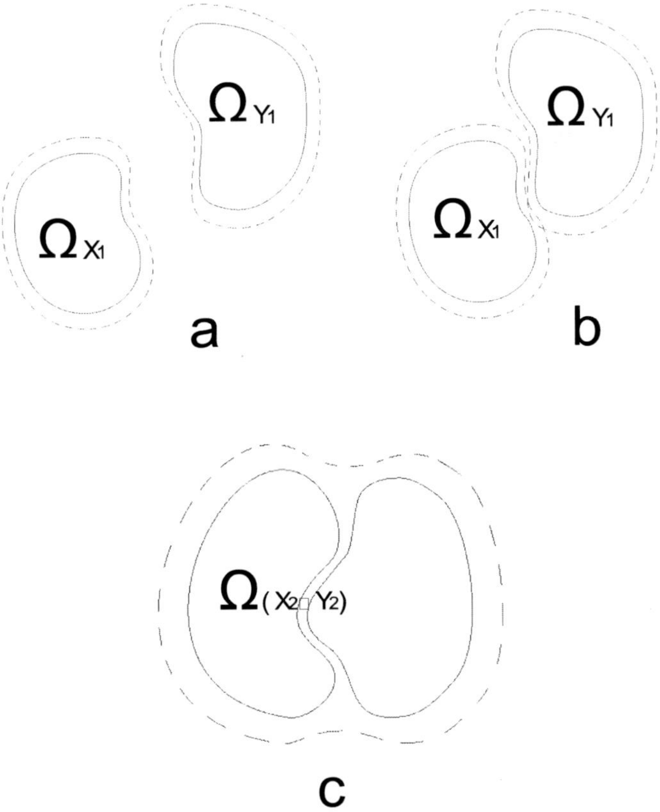

Figure 7. a. Two independent molecules; b. Invading each other; c. Forming new minimiser $(\mathbf{X}_2, \mathbf{Y}_2) \in \mathfrak{X}_{(\mathfrak{U}_1, \mathfrak{U}_2)}$.

Whenever one molecule invades another protein molecule's private domain, i.e., (42) is true, then this change to non-native structure shape phenomenon will happen. This can be used to explain many situations, such as protein docking, allosteric proteins, protein binding with other particles, and enzyme-substrate complex, etc., [7].

This can also be used to explain why protein stabilities are labile as described by Hien Wu in [44]. In a cell it is very crowded, there are a lot of chances that some particle invades a protein's private domain.

7.4 Denaturation and Refolding

Another phenomenon is that the physiological environment $\mathfrak{E}_N$ itself changed to a new environment $\mathfrak{E}_D$, then denaturation happens since the native structure $\mathbf{X}_N$ of the molecule $\mathfrak{U}$ is no longer a minimiser of the Gibbs free energy function $G(\mathbf{X}; \mathfrak{E}_D, \mathfrak{U})$, and usually, $g_{\mathfrak{E}_D,\mathfrak{U}}(G)$ is rugged, showing proteins evolved in the environment similar to $\mathfrak{E}_N$.

Besides explain protein folding and denaturation uniformly, study the rate of changing Gibbs free energy function with environment has other advantages. For example, if we only change the temperature T of the physiological environment $\mathfrak{E}_N$, let temperature $T(t)$ be a function of time, then we can get

$$\left.\frac{\mathrm{d}G(\mathbf{X}_N; T(t), \mathfrak{U})}{\mathrm{d}t}\right|_{t=0} = \left.\frac{\partial G(\mathbf{X}_N; T(t), \mathfrak{U})}{\partial T}\frac{\mathrm{d}T(t)}{\mathrm{d}t}\right|_{t=0} = -ST'(0),$$

where by (12), S is the entropy of the system $\mathfrak{S}_{\mathbf{X}_N}$. Thus if we lower temperature, G will increase at that moment; and if we increase the temperature, G will decrease at that moment. And the rate of changing temperature affecting the rate of the Gibbs free energy change. Similar trick can be applied to the pressure P.

If one designs a peptide chain to make a new protein like molecule, the Gibbs free energy functions under various environments will help to estimate in each environment what is the native structure (which is desired) will be, and can also help to control environments to make the folding to the desired native structure, one does not need imitating nature to make folding happens in a fixed environment.

8. Testable Predictions

To be useful, a scientific theory has to allow falsifying, i.e., has to make testable predictions. Now we make two predictions. One is the protein structure prediction by first principle; another is predicting that the origins of amyloid related and prion diseases are caused by environment changes, that is, these mis-folded proteins are denatured proteins.

8.1 First Principle Prediction of Protein Structure

Let $\mathfrak{U}$ be a globular protein and $\mathbf{X}_N$ its native structure. By the thermodynamic principle, the native structure $\mathbf{X}_N$ satisfies that

$$G(\mathbf{X}_N; \mathfrak{E}_N, \mathfrak{U}) = \min_{\mathbf{X} \in \mathfrak{X}_{\mathfrak{U}}} G(\mathbf{X}; \mathfrak{E}_N, \mathfrak{U}).$$

Thus, predicting what is $\mathbf{X}_N$ becomes a pure mathematical problem, minimisation of the Gibbs free energy function. In the minimisation process, we do not need any data base information. Besides, since we can derive the Gibbs free energy function $G(\cdot\,; \mathfrak{E}_N, \mathfrak{U}) : \mathfrak{X}_{\mathfrak{U}} \to \mathbb{R}$ from only the knowledge of $\mathfrak{U}$'s peptide chain, this structure prediction is truly in first principle, or *ab initio*.

When solving a minimisation problem, the first question is, are there minimisers? This is the existence problem. For our task, Theorem 6.4 guarantees that the existence of a minimiser of $G(\mathbf{X}; \mathfrak{E}_N, \mathfrak{U})$.

Next is how to find minimisers. As a mathematical problem, all mathematical tools can be applied, without any prohibitions such as one has to deal with when predicting native structures by molecular dynamics, no energy barriers, no need to keep to apply integrators to keep in a trajectory, no worry about energy traps, no longer depend on the ergodic theorems, and moreover, no need to construct force fields.

For example, if we have gradient (smoothness), an interior minimiser $\mathbf{X}_M$ of $G(\mathbf{X};$
$\mathfrak{E}_N, \mathfrak{U})$, global or local, have a common characteristic:

$$\nabla_{\mathbf{X}} G(\mathbf{X}_M; \mathfrak{E}_N, \mathfrak{U}) = \mathbf{0} = (\overbrace{0, \cdots, 0}^{3n}). \tag{44}$$

Minimisation of $G(\mathbf{X}; \mathfrak{E}_N, \mathfrak{U})$ can be performed by the fastest descending method, or Newton's method, as follows: Let $\mathbf{X}_0$ be an interior conformation and $s > 0$ be a small number, generating a sequence of conformations $\mathbf{X}_i$, $i = 1, \cdots, N$, by

$$\mathbf{X}_{i+1} = \mathbf{X}_i - s\nabla_{\mathbf{X}} G(\mathbf{X}_i; \mathfrak{E}_N, \mathfrak{U}), \quad i = 0, 1, \cdots, N-1.$$

By selecting s small enough, it is always true that either

$$G(\mathbf{X}_{i+1}; \mathfrak{E}_N, \mathfrak{U}) < G(\mathbf{X}_i; \mathfrak{E}_N, \mathfrak{U}),$$

or (44) is true for $\mathbf{X}_i$.

There will be limit $\mathbf{X}_M = \lim_{i\to\infty} \mathbf{X}_i$, such that either (44) is true, $\mathbf{X}_M$ is an interior local minimiser, or it is a boundary minimiser. This is the fastest descending method, or Newton's method.

8.2 Combined Fastest Descending Minimisation

We first solve the minimisation problem in an Equi-AD class $[\mathbf{X}]$ for an interior conformation $\mathbf{X}$. If

$$\nabla_\Phi G_{[\mathbf{X}]}(\Phi(\mathbf{X}); \mathfrak{E}_N, \mathfrak{U}) \neq \mathbf{0},$$

then let

$$M = \max\{|\delta_{L_1} G_{[\mathbf{X}]}(\Phi; \mathfrak{E}_N, \mathfrak{U})|, \cdots, |\delta_{L_L} G_{[\mathbf{X}]}(\Phi; \mathfrak{E}_N, \mathfrak{U})|\},$$

and

$$\theta_i = \frac{2\pi}{M} \delta_{L_i} G_{[\mathbf{X}]}(\Phi; \mathfrak{E}_N, \mathfrak{U}), \quad \Theta = (\theta_1, \cdots, \theta_L).$$

Take a small $s > 0$, the total transformer rotation $\mathbf{X}_s = \mathfrak{R}(-s\Theta)(\mathbf{X})$ then changes the coordinates $\Phi(\mathbf{X}) = (\phi_1, \cdots, \phi_L)$ to $\Phi(\mathbf{X}_s) = (\phi_1 - s\theta_1, \cdots, \phi_L - s\theta_L)$. Moreover, for s small enough,

$$G_{[\mathbf{X}]}(\Phi(\mathbf{X}_s); \mathfrak{E}_N, \mathfrak{U}) < G_{[\mathbf{X}]}(\Phi(\mathbf{X}); \mathfrak{E}_N, \mathfrak{U}).$$

The good point is, if $\mathbf{X} \notin \partial\mathfrak{X}_{\mathfrak{U}}$, then $[\mathbf{X}] \cap \partial\mathfrak{X}_{\mathfrak{U}} = \emptyset$. Thus, repeat this process, we will arrive at a $\mathbf{X}_M$ with $\Phi_M = \Phi(\mathbf{X}_M)$, such that

$$\nabla_\Phi G_{[\mathbf{X}]}(\Phi_M; \mathfrak{E}_N, \mathfrak{U}) = \mathbf{0} = (\overbrace{0, \cdots, 0}^{L}),$$

so we find an interior minimiser of $G_{[\mathbf{X}]}$. Afore mentioned minimisations in [15] used this method.

Note that the above $\mathbf{X}_M$ is not necessarily a minimiser of G because $[\mathbf{X}]$ is only a lower dimensional subset of $\mathfrak{X}_{\mathfrak{U}}$. To get a minimiser of G, we start with an interior conformation $\mathbf{X}_0 \in \mathfrak{X}^\circ_{\mathfrak{U}}$, get above minimiser $\mathbf{X}_M \in \mathfrak{X}^\circ_{\mathfrak{U}}$, of $G_{[\mathbf{X}_0]}$, then if

$$\nabla_{\mathbf{X}} G(\mathbf{X}_M; \mathfrak{E}_N, \mathfrak{U}) \neq \mathbf{0},$$

we set $\mathbf{X}_1 \in \mathfrak{X}^\circ_{\mathfrak{U}}$, $\mathbf{X}_1 \notin [\mathbf{X}_0]$, such that $\mathbf{X}_1 = \mathbf{X}_M - s\nabla_{\mathbf{X}} G(\mathbf{X}_M; \mathfrak{E}_N, \mathfrak{U})$, for $s > 0$ small enough, it must be

$$G(\mathbf{X}_1; \mathfrak{E}_N, \mathfrak{U}) < G(\mathbf{X}_M; \mathfrak{E}_N, \mathfrak{U}).$$

Then we take minimisation in $[\mathbf{X}_1]$. Repeat this process, alternatively doing fastest descending in Φ and $\mathbf{X}$ variables, we always operating in the interior of $\mathfrak{X}_{\mathfrak{U}}$, until we achieve a minimiser of G, but it may still be a boundary minimiser.

8.3 Amyloids and Prions

8.3.1 Amyloids

Amyloids are insoluble fibrous protein aggregates sharing specific structural traits [33]. There are a number of diseases in which normally soluble proteins associate into regular, insoluble amyloid fibrils [34]. Besides in disease cases, amyloids also appear in organisms as normally functioning proteins [33].

It is reported that chameleon sequences are often associated with amyloids [17]. A chameleon sequence's secondary structure in a amyloid is different from its secondary structure in the native structure of the protein. Since the whole amino acid sequence does not change, the difference between an amyloid and a native structure can only be caused by environment change, or initial conformation change.

If no environment change, then amyloid and native structure are different minimisers of the same Gibbs free energy function, implying that amyloids and normal proteins fold from different initial conformations contained in different maximum converging domains. Given normal proteins have almost all initial conformations concentrated in one maximum converging domain $Q_{\mathbf{X}_N}$, where $\mathbf{X}_N$ is the native structure, why amyloids taking different initial conformations to another maximum converging domain while keeping in the same environment? The only answer is that the environment really has been changed. Let $\mathfrak{E}_A$ be the amyloid environment.

Write an amyloid conformation as $\mathbf{X}_A$ and native structure as X_N, then the explanation is:

$$G(\mathbf{X}_N; \mathfrak{E}_N, \mathfrak{U}) < G(\mathbf{X}_A, \mathfrak{E}_N, \mathfrak{U}); \quad G(\mathbf{X}_A; \mathfrak{E}_A, \mathfrak{U}) < G(\mathbf{X}_N; \mathfrak{E}_A, \mathfrak{U}).$$

Different functions have different minimisers, it is so plain. The single molecule theory of protein folding then predicts that the cause of amyloids related diseases is changed local environment. It is just denaturation of proteins, but happen in living organism such that instead of taking various minimisers, taking just one minimiser $\mathbf{X}_A$. In this environment $\mathfrak{E}_A$, these new minimiser conformations of different molecules invade each other's private domain to form complex, creating aggregate environments that makes the amyloid a minimiser in corresponding Gibbs free energy function.

8.3.2 Prions

A prion is a mis-folded protein that may cause disease. In fact, the protein has different natives structures in different environments, disease causing prions must be environment favour prion structures replaced the normal physiological environment. Prion proteins are unprecedented infectious pathogens, prion proteins enter an organism may cause the prion proteins in normal native structures changing to the disease native structures, [32]. Like amyloids, the mechanism it that first large quantities of prion protein molecules enter a cell to cause an environment change. In the new environment, a prion protein molecule and normal protein molecule being able to invade each other's private domain to form new immediate environment, both of them taking the prion native structure is a minimiser of the corresponding Gibbs free energy function, the disease being spread.

9. Conclusion

Toward a single molecule theory of protein folding, a Gibbs free energy function $G(\cdot\,; \mathfrak{C}_N, \mathfrak{U}) : \mathfrak{X}_{\mathfrak{U}} \to \mathbb{R}$ is established. It is lower semi-continuous, and has first derivatives away from the set G is infinity. Holistic views about protein structure and protein folding can be drawn from the derived Gibbs free energy function.

Many natural phenomena in protein folding and protein structures can be explained with this Gibbs free energy function and its dependence on environment. Among them are a holistic view of protein structure which explains phenomenon such as the chameleon sequences; why just reducing hydrophobic surface area can produce secondary structures and hydrogen bonds; structure stability; allosteric proteins, docking, enzyme-substrate combination; and denaturation and refolding. The new unified explanations are at least as good as current, *ad hoc* ones, in some cases are much better.

Testable predictions are given, first among them is the prediction of protein native structure from the only knowledge of the protein's amino acid sequence. It just becomes a pure mathematical problem of minimisation. Another testable prediction is that changed local environment causes amyloid and prion related diseases.

Acknowledgment:

The author appreciate Yupeng Liu for drawing Figures 6 and 7.

References

References

[1] Anfinsen, C. B., 1973. Principles that govern the folding of protein chains. *Science* 181, 223-230.

[2] Anson, M. L. and Mirsky, A. E., 1925. On some general properties of proteins. *Journal of General Physiology*, 9, 169-179.

[3] Bader, R. F. W. *Atoms in Molecules: A Quantum Theory.* Clarendon Press · Oxford, 1990.

[4] Bahar, I., Kaplan, M., and Jernigan, R. L., Short-range conformational energies, secondary structure propensities, and recognition of correct sequence-structure matches. *Proteins: Structure, Function, and Genetics*, 29: 292-308, 1997.

[5] Ben-Naim, A., Pitfalls in Anfinsen's thermodynamic hypothesis. *Chemical Physics Letters* 511, 126-128, 2011.

[6] Born, M. and Oppenheimer J. R., On the quantum theory of molecules. *Annalen der Physik*, (in German) 389(80), 457-484, 1927.

[7] Branden, C. and Tooze, J., *Introduction to Protein Structure* Second Edition, Garland Publishing Inc., New York, 1999.

[8] Connolly, M. L. Analytical molecular surface calculation. *J. Appl. Cryst.*, 16:548-558, 1983.

[9] Dai, X., 2007. *Advanced Statistical Physics.* Shanghai, Fudan University Press.

[10] Dill, K. A. Dominant forces in protein folding. *Biochemistry*, 29 7133-7155, 1990.

[11] Dunker, A. K., et all. Intrinsically disordered protein. *Journal of Molecular Graphics and Modelling*, 19(1): 26-59, 2001.

[12] E. Eisenberg and A. D. McLachlan, *Nature*, 319, 199 (1986).

[13] Fang, Y., 2005. Mathematical protein folding problem. In: Hoffman, D. (ed.), *Global Theory of Minimal Surfaces. Proceedings of the Clay Mathematical Proceedings*, Cambridge Massachusetts, American Mathematical Society, Clay Mathematics Institute, pp. 611-622.

[14] Fang, Y. and Jing, J. Implementation of a mathematical protein folding model. *International Journal of Pure and Applied Mathematics* 42(4), 481-488, 2008.

[15] Fang, Y. and Jing, J. Geometry, thermodynamics, and protein. *Journal of Theoretical Biology*, 262, 382-390, 2010.

[16] Fang, Y. Gibbs free energy formula for protein folding. In: Morales-Rodriguez, R. (ed.), *Thermodynamics - Fundamentals and Its Application in Science*. pp. 47-82. DOI: 10.5772/2615. Open source: http://www.intechopen.com/books/thermodynamics-fundamentals-and-its-application-in-science

[17] Guo, J., Jaromczyk, J. W., and Xu, Y., Analysis of chameleon sequences and their implications in biological processes. *Proteins: Structure, Function, and Bioinformatics*. 67: 548-558, 2007.

[18] Haynes, P. D., 1998. Linear-scaling methods in *ab initio* quantum-mechanical calculations. A dissertation submitted for the degree of Doctor of Philosophy at the University of Cambridge. (1998).

[19] Hubner, I. A., Shakhnovic, E. I. Geometric and physical considerations for realistic protein models. *Phys. Rev. E* 72, 022901. DOI: 10.1103/PhysRevE.72.022901, 2005.

[20] Karplus, M., Behind the folding funnel diagram. *Nature Chemical Biology* 7: 401-403, 2011.

[21] Kauzmann, W. Some factors in the interpretation of protein denaturation. *Adv. Protein Chem.* 14, 1-63, 1959.

[22] Le Bellac, M., Mortessagne, F., and Batrouni, G. G., 2004. *Equilirium and Non-Equilibrium Statistical Thermodynamics.* Cambridge University Press, Cambridge.

[23] Lee, B. and Richards, F. M. The interpretation of protein structures: estimation of static accessibility. *J. Mol. Biol.* 55:379-400, 1971.

[24] Levinthal, C. Are there pathways for protein folding. *Journal de Chimie Physique* 65(1):44-45, 1968.

[25] Levinthal, C. How to fold graciously. Mössbauer Spectroscopy in Biological Systems Proceedings. Proceedings of a meeting held at Allerton House, Monticello, Illinois. *University of Illinois Bulletin* 67(41): 22-26, 1969. http://www-miller.ch.cam.ac.uk/levinthal/levinthal.html

[26] Mann, G., 1906. *Chemistry of Proteins.* London, Macmillan.

[27] Mirsky, A. E. and Pauling, L. On the structure of native, denatured, and coagulated proteins. *Proceedings of the National Academy of Science USA* 22(7): 439-447, 1936.

[28] Myatt, C. J., King, B. E., Turchette, Q. A., Sackett, C. A., Kielinski, D., Itano, W. M., Monore, C., and Wineland, D. J. Decoherence of quantum superpositions through coupling to engineered reservoirs. *Nature* 403, 269-273, 2000.

[29] Novotny, J., Bruccoleri, R., and Karplus, M. An analysis of incorrectly folded protein models. Implications for structure predictions. *J. Mol. Biol.*, 177:787-818, 1984.

[30] Novotny, J., Rashin, A. A., and Bruccoleri, R. Criteria that discriminate between native proteins and incorrectly folded models. *Proteins*, 4:19-30, 1988.

[31] Popelier, P. *Atoms in Molecules: An Introduction* Prentice Hall, 2000.

[32] Prusiner, S. B., Prions. *PNAS* 95(23): 13363-13383, 1998.

[33] Pulawsky, W., Ghoshdastider, U., Andrisano, V., and Fillipek, S. Ubiquitous Amyloids. *Appl. Biochem. Biotechnol.* 166(7), 1626-1643, 2012.

[34] Ramirez-Alvarado, M., Merkel, J. S., and Regan, L., A systematic exploration of the influence of the protein stability on amyloid fibril formation in vitro. *PNAS* 97(16): 8979-8984, 2000.

[35] Richards, F. M. Areas, volumes, packing, and protein structure. *Ann. Rev. Biophys. Bioeng.* 6 151-176, 1977.

[36] Rose, G. D., Fleming, P. J., Banavar, J. R., Maritan, A. A backbone based theory of protein folding. *PNAS* 103(45), 16623-16633, 2006.

[37] Rouvray, D. Model answers? *Chemistry in Britain* 35(9), 30-32, 1999.

[38] Rouvray, D. Atoms as hard spheres. *Chemistry in Britain* 36(5), 25, 2000.

[39] Schafer. N. P., Kim, B. L., Zheng, W., and Wolynes, P. G., Learning to fold proteins using energy landscape theory. *Israel Journal of Chemistry*, 54(8-9), 1311-1337, 2014.

[40] Schrödingder, E. *Statistical Thermodynamics.* Dover Publications Inc. New York, 1989.

[41] Serdyuk, I. N., Zaccai, N. R., and Zaccai, J., *Methods in Molecular Biophysics: Structure, Dynamics, Function.* Cambridge University Press, Cambrigde, 2007.

[42] Simon, L, Lectures on Geometric Measure Theory. Proceedings of the CMA, ANU, Canberra, 1983.

[43] Tanford, C. and Reynolds, J., 2001 *Nature's Robots: A History of Proteins.* Oxford, University Press.

[44] Wu, H., 1929. A theory of denatured and coagulated proteins. *America Journal of Physiology*, 90, 562-563.

[45] Wu, H., 1931. Studies of denaturation of proteins XIII. A theory of denaturation. *Chinese J. Physiol.*, 5 321-344, 1931. Republished in *Advances in Protein Chemistry*, 46 6-26, 1995.

[46] Yin, H., Landick, R., and Gelles, J., Tethered particle motion method for studying transcript elongation by a single RNA polymerase molecule. Biophys J., 67: 2468-2478, 1994.

In: Advances in Protein Folding Research ISBN: 978-1-63483-593-0
Editor: Mathew Hale

BIBLIOGRAPHY

Biochemical pathways: An atlas of biochemistry and molecular biology, LCCN: 2011041441, Main title: Biochemical pathways: an atlas of biochemistry and molecular biology / edited by Dietmar Schomburg, Gerhard Michal. Edition: 2nd ed. Published/Created: Hoboken, N.J.: John Wiley & Sons, c2012. Description: xi, 398 p.: ill. (some col.); 29 cm. Links: Publisher description http://www.loc.gov/catdir/enhancements /fy1203/2011041441-d.html; Table of contents only http://www.loc.gov /catdir/enhancements/fy1203/2011041441-t.html, Contributor biographical information http://www.loc.gov/catdir/enhancements/fy1212/2011041441-b.html, ISBN: 9780470146842, LC classification: QP171. B685 2012, Related names: Schomburg, D. (Dietmar) Michal, Gerhard. Summary: "Covering a wide range of subject matter, including biochemistry, molecular and cell biology, medicine, chemistry, and allied health, Biochemical Pathways is a full-color, easy-to-use resource for students and professionals. This information-packed reference features a unique summary of biochemical pathways based on the well-known Biochemical Pathways chart. Included is descriptive information about properties such as enzymes, chemicals, proteins, and DNA, all of which act together to create an elaborate chain that drives all biological functions. Completely updated, this new edition continues to play a valuable role in this important scientific field" - Provided by publisher. "Biochemical Pathways: An Atlas of Biochemistry and Molecular Biology, Second Edition, explains the multi-step chemical reactions (for example, carbohydrate metabolism pathways) that occur in biological systems (e.g., within cells and organelles). The chemical reactions, location within the cell, structure and function of proteins and nucleic acids (including the enzymes), small molecules (carbohydrates, amino acids, lipids, steroids, nucleotides) involved, vitamins and cofactors present, storage of information in DNA and translation into proteins, viruses infecting systems, various protein mechanisms, respiration processes, cellular communication and regulation, transport, defense, and blood coagulation are all discussed in detail" - Provided by publisher. Contents: Machine generated contents note: Chapter 1. Introduction and general aspects, chemistry and physical chemistry Gerhard Michael and Dietmar Schomburg, Chapter 2. The cell and its contents: Enzymes, nucleic acids, and polymeric carbohydrates and lipids Gerhard Michal and Dietmar Schomburg, Chapter 3. General metabolism in animals, plants and bacteria 3.1. Carbohydrate Metabolism and Citrate Cycle, 3.2. Amino Acids and Derivatives, 3.3. Tetrapyrroles, 3.4. Lipids and Glycolipids, 3.5. Steriods and Isopreniods, 3.6. Nucleotides and Nucleosides, 3.7. Cofactors and Vitamins, 3.8. Nucleic Acid Metabolism in Bacteria, 3.9. Nucleic Acid Metabolism in Eukarya, 3.10. Special Bacterial Metabolism and Biosynthesis of Antimicrobials, 3.11. Electron Transfer Reactions and Oxidative Phosphorylation, 3.12. Photosynthesis, 3.13. Plant Secondary Metabolism, Chapter 4. Protein biosynthesis, modification and degradation, 4.1. Protein Synthesis in Bacteria, 4.2. Protein Biosynthesis in Eukarya, 4.3. Cell Cycle in Eukarya, 4.4. Posttranslational Modification of Proteins, 4.5. Protein Folding, Transport / Targeting and Degradation, Chapter 5. Viruses Klaus Klumpp, Chapter 6. Transport systems, 6.1. Transport through Membranes, 6.2. Transport of Lipids in Plasma, 6.3. Oxygen Transport by Hemoglobin Chapter 7. Signal transduction and cellular communication Gerhard Niederfellner, Chapter 8. Immune system, 8.1. Components of the Immune System, 8.2. Generation of a Specific Immune Response, 8.3. Pathologic Immune responses, 8.4. Adhesion of Leukocytes, Chapter 9. Blood coagulation and fibrinolysis, P. Mueller, Chapter 10. Biochemical networks, bioinformatics and systems biology, Dietmar Schomburg. Subjects: Metabolism – Atlases. SCIENCE / Life Sciences / Biochemistry. Notes: Includes bibliographical references and index.

Biomolecular simulations: *Methods and protocols*, LCCN: 2012946729. Main title: Biomolecular simulations: Methods and protocols / edited by Luca Monticelli, Emppu Salonen. Published/Created: New York: Humana Press; Springer, c2013. Description: xiii, 702 p.: ill. (some col.); 26 cm. Links: Publisher description http://www.loc.gov/catdir/enhancements /fy1306/2012946729-d.html. Table of contents only http://www.loc.gov

/catdir/enhancements/fy1306/2012946729-t.html, ISBN: 9781627030168 (hbk.: alk. paper) 1627030166 (hbk.: alk. paper) LC classification: QP517.M65 B55 2013, Related names: Monticelli, Luca edt, Salonen, Emppu edt, Contents: Ab Initio, density functional theory, and semi-empirical calculations / Mikael P. Johansson, Ville R.I. Kaila, and Dage Sundholm - Ab Initio molecular dynamics / Kari Laasonen -- Introduction to QM/MM simulations / Gerrit Groenhof - Computational enzymology / Alessio Lodola and Adrian J. Mulholland - QM and QM/MM simulations of proteins / Thomas Steinbrecher and Marcus Elstner -- Classical molecular dynamics in a nutshell / Susanna Hug - Enhanced sampling algorithms / Ayori Mitsutake, Yoshiharu Mori, and Yuko Okamoto - Force fields for classical molecular dynamics / Luca Monticelli and D. Peter Tieleman -- Polarizable force fields / Hanne S. Antila and Emppu Salonen - Electrostatics interactions in classical simulations / G. Andres Cisneros, Volodymyr Babin, and Celeste Sagui - Introduction to best practices in free energy calculations / Michael R. Shirts and David L. Mobley - Recipes for free energy calculations in biomolecular systems / Mahmoud Moradi ... [et al.] - Molecular docking methodologies / Andrea Bortolato ... [et al.] - Simulation studies of the mechanism of membrane transporters / Giray Enkavi ... [et al.] - Molecular dynamics simulations of lipid bilayers: simple recipe of how to do it / Hector Martinez-Seara and Tomasz Rog - Simulations of lipid monolayers / Svetlana Baoukina and D. Peter Tieleman - Simulating DNA by molecular dynamics: aims, methods, and validation / Nicolas Foloppe, Marc Gueroult, and Brigitte Hartmann - Simulation of carbohydrates, from molecular docking to dynamics in water / Nicolas Sapay, Alessandra Nurisso, and Anne Imberty - Systematic methods for structurally consistent coarse-grained models / W.G. Noid - Martini coarse-grained force field / Xavier Periole and Siewert-Jan Marrink - Multiscale molecular modeling / Matej Praprotnik and Luigi Delle Site - Coarse-grained models for protein folding and aggregation / Philippe Derreumaux - Elastic network models: theoretical and empirical foundations / Yves-Henri Sanejouand - Introduction to dissipative particle dynamics / Zhong-Yuan Lu and Yong-Lei Wang -- Multiscale molecular dynamics simulations of membrane proteins / Syma Khalid and Peter J. Bond - Vesicles and vesicle fusion: coarse-grained simulations / Julian C. Shillcock. Subjects: Molecular dynamics--Mathematical models. Biomolecules - Mathematical models. Molecular biology - Research. Molecular biology - Computer simulation. Molecular Dynamics Simulation Models, Molecular Computer Simulation

Form/Genre: Laboratory Manuals Notes: Includes bibliographical references and index. Series: Methods in molecular biology; 924Methods in molecular biology (Clifton, N.J.); v. 924.

Bio-nanoimaging: protein misfolding & aggregation, LCCN: 2013444579 Main title: Bio-nanoimaging: protein misfolding & aggregation / edited by Vladimir N. Uversky, Associate Professor, University of South Florida, College of Medicine, Department of Molecular Medicine, Miami, FL, USA, Yuri L. Lyubchenko, Professor of Pharmaceutical Sciences, Director, Nanoimaging Core Facility, College of Pharmacy, University of Nebraska Medical Center, Omaha, NE, USA. Published/Produced: Amsterdam; Boston: Elsevier/AP, Academic Press is an imprint of Elsevier, [2014] ©2014 Description: xix, 526 pages: illustrations (some color); 29 cm ISBN: 97801239443130123944317, LC classification: QP551 .B4675 2014, Related names: Uversky, Vladimir N., editor of compilation. Lyubchenko, Yuri L., editor of compilation. Subjects: Protein folding--Computer simulation. Imaging systems. Nanotechnology. Protein Folding. Amyloid. Cell Aggregation. Molecular Diagnostic Techniques. Nanotechnology. Proteins. Proteostasis Deficiencies. Notes: Includes bibliographical references and index. Additional formats: Online version: Bio-nanoimaging. London: Academic Press/Elsevier, [2014] 9780123978219 (OCoLC) 862614099.

Biophysics of RNA folding, LCCN: 2012952011, Main title: Biophysics of RNA folding / Rick Russell, editor. Published/Created: New York, NY: Springer, c2013. Description: vi, 236 p.: ill. (some col).; 24 cm. Links: Publisher description, http://www.loc.gov/catdir/enhancements/fy1307/2012952011-d.html, Table of contents only http://www.loc.gov/catdir/enhancements/fy1307/2012952011-t.html, ISBN: 97814614495391461449537, LC classification: QP623. B56 2013, Related names: Russell, Rick. Contents: Introduction / Rick Russell - Comparative Analysis of the Higher-order Structure of RNA / Robin R. Gutell - Graph Applications to RNA Structure and Function / Namhee Kim, Katherine Niccole Fuhr, Tamar Schlick - Prediction and Coarse-Grained Modeling of RNA Structures / Zhen Xia, Pengyu Ren - Studying RNA Folding Using Site-Directed Spin Labeling / Xiaojun Zhang, Peter Z. Qin - The RNA Recognition Motif and Messenger RNA / Kathleen B. Hall - Memory Effects in RNA folding dynamics revealed by single molecule fluorescence / Rui Zhao, David Rueda - An integrated picture of HDV ribozyme catalysis / Barbara L. Golden, Sharon Hammes-Schiffer, Paul R. Carey, Philip C. Bevilacqua -- Combining biochemical and structural information to model RNA-protein complex assembly / Maithili Saoji, Chun Geng, Paul J. Paukstelis -- Following RNA Folding From Local and Global Perspectives / Michael Brenowitz, Lois Pollack -- The Roles of Chaperones in RNA Folding / Rick Russell. Subjects: RNA. Biophysics. Protein folding. Notes: Includes bibliographical references and index. Series: Biophysics for the life sciences.

Energy landscapes, inherent structures, and condensed-matter phenomena, LCCN: 2015004169, Personal name: Stillinger, F. H., author. Main title: Energy landscapes, inherent structures, and condensed-matter phenomena / Frank H. Stillinger. Published/Produced: Princeton, New Jersey: Princeton University Press, [2015] ©2015 Projected pub date: 1512 Description: pages cm, ISBN: 9780691166803 (hardback: alk. paper) LC classification: QC173.457.S7 S75 2015, Contents: Potential energy functions -- Statistical mechanical basics - Basins, saddles, and configuration-space mapping -- Crystal phases - Liquids at thermal equilibrium -- Supercooled liquids and glasses - Low-density matter - The helium isotopes - Water - Polymeric substances - Protein folding phenomena. Subjects: Condensed matter. Nuclear physics. Notes: Includes bibliographical references and index.

Fuzziness: Structural disorder in protein complexes, LCCN: 2011021158, Main title: Fuzziness: structural disorder in protein complexes / edited by Monika Fuxreiter, Peter Tompa. Published/Created: Austin, TX: Springer Science+Business Media; New York: Landes Bioscience, c2012. Description: xx, 194 p.: ill.; 27 cm. Links: Contributor biographical information http://www.loc.gov/catdir/enhancements/fy1316/2011021158 -b.html. Publisher description http://www.loc.gov/catdir/enhancements /fy1316/2011021158-d.html. Table of contents only http://www.loc.gov /catdir/enhancements/fy1316/2011021158-t.html ISBN: 9781461406587 LC classification: QP551 .F99 2012, Related names: Fuxreiter, Monika, 1969-Tompa, Peter. Contents: Fuzzy complexes: a more stochastic view of protein function / Monika Fuxreiter and Peter Tompa -- Dynamic fuzziness during linker histone action / Steven J. McBryant and Jeffrey C. Hansen -- Intrinsic protein flexibility in regulation of cell proliferation: advantages for signaling and opportunities for novel therapeutics / Ariele Viacava Follis, Charles A. Galea, and Richard W. Kriwacki -- Interplay between protein order, disorder and oligomericity in receptor signaling / Alexander B. Sigalov -- Consequences of fuzziness in the NFkB/IkBa interaction / Elizabeth A. Komives -- Roles for intrinsic disorder and fuzziness in generating context specific function in ultrabithorax: a hox transcription factor / Sarah E. Bondos and Hao Ching Hsiao -- Molecular recognition by the EWS transcriptional activation domain / Kevin A.W. Lee -- The measles virus NTAILXD complex: an illustrative example of fuzziness / Sonia Longhi -- Fuzziness in the core of the human pathogenic viruses HCV and HIV / Roland Ivanyi Nagy and Jean Luc Darlix -- Structural disorder and protein elasticity / Sarah Rauscher and Røgis Pomès -- "Fuzziness" in the cellular interactome: a historical perspective / G. Rickey Welch. Subjects: Proteins--Conformation. Protein folding. Protein binding. Fuzzy systems. Protein Conformation. Protein Denaturation. Notes: Includes bibliographical references and index. Series: Advances in experimental medicine and biology; no. 725.

Membrane structural biology: With biochemical and biophysical foundations, LCCN: 2013028009, Personal name: Luckey, Mary, 1948- Main title: Membrane structural biology: with biochemical and biophysical foundations / Mary Luckey. Edition: Second edition. Published/Produced: Cambridge; New York: Cambridge University Press, 2014. Description: xiv, 411 pages: color illustrations; 28 cm ISBN: 9781107030633 (hardback) LC classification: QH601. L75 2014, Summary: "This textbook provides a strong foundation and a clear overview for students of membrane biology and an invaluable synthesis of cutting edge research for working scientists. The text retains its clear and engaging style, providing a solid background in membrane biochemistry, while also incorporating the approaches of biophysics, genetics, and cell biology to investigations of membrane structure, function, and biogenesis to provide a unique overview of this fast-moving field. A wealth of new high resolution structures of membrane proteins are presented, including the Na/K pump and a receptor-G protein complex, offering exciting insights into how they function. All key tools of current membrane research are described, including detergents and model systems, bioinformatics, protein-folding methodology, crystallography and diffraction, and molecular modeling. This comprehensive and up-to-date text, emphasizing the correlations between membrane research and human health, provides a solid foundation for all those working in this field"-- Provided by publisher. Contents: The diversity of membrane lipids -- Tools for studying membrane components -- Proteins in or at the bilayer -- Bundles and barrels -- Functions and families -- Protein folding and biogenesis -- Diffraction and simulation -- Membrane enzymes -- Membrane receptors -- Transporters -- Channels -- Electron transport and energy transduction -- In pursuit of complexity. Subjects: Membranes

(Biology) Membrane lipids. Membrane proteins. SCIENCE / Life Sciences / Biology / Molecular Biology. Notes: Includes bibliographical references and index.

Methods in protein biochemistry, LCCN: 2011029558, Main title: Methods in protein biochemistry / edited by Harald Tschesche. Published/Created: Berlin: De Gruyter, c2012. Description: xxiii, 352 p.: ill.; 25 cm. ISBN: 9783110252330 (alk. paper) 3110252333 (alk. paper) 9783110252361 (e-ISBN) 3110252368 (e-ISBN) LC classification: QP551 .M3883 2012 Related names: Tschesche, Harald. Contents: Three-phase partitioning -- Folding and degradation functions of molecular chaperones -- Membrane protein folding in detergents -- Glycoprotein folding quality control in the endoplasmic reticulum -- Protein import into the intermembrane space of mitochondria -- On-membrane identification of gel-resolved proteins by MALDI-MS -- Analysis of protein complexes using chemical cross-linking and mass spectrometry -- Single-crystal spectroscopy correlated with X-ray crystallography provides complementary perspectives on macromolecular function -- Wide-angle X-ray solution scattering (WAXS) -- Where purity matters: recombinant versus synthetic peptides in beta amyloid formation -- Chemical modification of proteins in living cells -- Proteomics of human bronchoalveolar lavage fluid: discovery of biomarkers of chronic obstructive pulmonary disease with difference gel electrophoresis (DIGE) and mass spectrometry -- Proteomic of Duchenne muscular dystrophy -- Target-oriented peptide arrays in a palliative approach to cystic fibrosis -- Probing protein dynamics in vivo using backbone cyclization: bacterial acyl carrier protein as a case study -- The protein epitope mimetic approach to protein-protein interaction inhibitors -- The structural biology of alpha-1 antitrypsin deficiency and the serpinopathies. Subjects: Proteins--Analysis. Proteomics--Methodology. Proteins--analysis. Proteins--chemistry. Proteins--isolation & purification. Proteomics--methods. Notes: Includes bibliographical references and index.

Molecular chaperones, LCCN: 2012952678, Main title: Molecular chaperones / Sophie Jackson, editor; with contributions by A. Ahmad ... [et al.]. Published/Created: Heidelberg; New York: Springer Verlag, c2013. Description: ix, 272 p.: ill. (some col.); 24 cm. ISBN: 9783642345517 (alk. paper) 3642345514 (alk. paper) 9783642345524 (eBook) 3642345522 (eBook) LC classification: QP552.M64 M63 2013, Related names: Jackson, Sophie (Sophie E.)Ahmad, A. (Atta) Contents: Assisting oxidative protein folding: how do protein disulphide-isomerases couple conformational and chemical processes in protein folding? -- Peptide bond cis/trans isomerases: a biocatalysis perspective of conformational dynamics in proteins -- Small heat-shock protein: paramedics of the cell -- Allostery in the Hsp70 chaperone proteins -- Hsp90: structure and function -- Extracellular chaperones. Subjects: Molecular chaperones. Protein folding. Molecular Chaperones. Protein Folding. Notes: Includes bibliographical references and index. Additional formats: Also issued online. Series: Topics in current chemistry, 0340-1022; 328Topics in current chemistry; 328. 0340-1022.

Molecular genetics of bacteria, LCCN: 2012027461, Main title: Molecular genetics of bacteria / Larry Snyder ... [et al.]. Edition: 4th ed. Published/Created: Washington, DC: ASM Press, c2013. Description: xx, 707 p.: ill.; 29 cm. ISBN: 9781555816278 (hardcover: alk. paper) 1555816274 (hardcover) 9781555817169 (e-book) 1555817165 (e-book) LC classification: QH434. S59 2013, Related names: Snyder, Larry. Snyder, Larry. Molecular genetics of bacteria. Contents: The bacterial chromosome: DNA structure, replication, and segregation -- Bacterial gene expression: transcription, translation, and protein folding -- Bacterial genetic analysis: fundamentals and current approaches -- Plasmids -- Conjugation -- Transformation -- Lytic bacteriophages: development, genetics, and generalized transduction -- Lysogeny: the lambda paradigm and the role of lysogenic conversion in bacterial pathogenesis -- Transposition, site-specific recombination, and families of recombinases -- Molecular mechanisms of homologous recombination -- DNA repair and mutagenesis -- Regulation of gene expression: genes and operons -- Global regulation: regulons and stimulons -- Bacterial cell biology and development. Subjects: Bacteria--genetics. Bacteriophages--genetics. Chromosomes, Bacterial. Genetics, Microbial--methods. Molecular Biology--methods. Notes: Rev. ed. of: Molecular genetics of bacteria / Larry Snyder and Wendy Champness. c2007.Includes bibliographical references and index.

Molecular modeling of proteins, LCCN: 2014945209, Main title: Molecular modeling of proteins / edited by Andreas Kukol, University of Hertfordshire, Hatfield, Hertfordshire, UK. Edition: Second edition. Published/Produced: Totowa, NJ: Humana Press, [2015] ©2015, Description: x, 474 pages: illustrations (some color); 26 cm. ISBN: 9781493914647 (hbk: acid-free paper) 1493914642 (hbk: acid-free paper) LC classification: QP551. M624 2015, Related names: Kukol, Andreas. Contents: Molecular dynamics simulations / Erik R. Lindahl -- Monte

Carlo simulations / David J. Earl and Michael W. Deem -- Hybrid quantum and classical methods for computing kinetic isotope effects of chemical reactions in solutions and in enzymes / Jiali Gao ... [et al.] -- Comparison of protein force fields for molecular dynamics simulations / Olgun Guvench and Alexander D. MacKerell, Jr. -- Normal modes and essential dynamics / Steven Hayward and Bert L. de Groot -- Calculation of absolute protein-ligand binding constants with the molecular dynamics free energy perturbation method / Hyung-June Woo -- Free energy calculations applied to membrane proteins / Christophe Chipot -- Molecular dynamics simulations of membrane proteins / Philip C. Biggin and Peter J. Bond -- Membrane-associated proteins and peptides / Marc F. Lensink -- Implicit membrane models for membrane protein simulation / Michael Feig -- Comparative modeling of proteins / Gerald H. Lushington -- Transmembrane protein models based on high-throughput molecular dynamics simulations with experimental constratints / Andrew J. Beevers and Andreas Kukol -- Nuclear magnetic resonance-based modeling and refinement of protein three-dimensional structures and their complexes / Gloria Fuentes, Aalt D.J. van Dijk, and Alexandre M.J.J. Bonvin -- Conformational changes in protein function / Haiguang Liu ... [et al.] -- Protein folding and unfolding by all-atom molecular dynamics simulations / Hongxin Lei and Yong Duan -- Modeling of protein misfolding in disease / Edyta B. Małolepsza -- Identifying putative drug targets and potential drug leads: starting points for virtual screening and docking / David S. Wishart -- Receptor flexibility for large-scale in silico ligand screens: chances and challenges / B. Fischer, H. Merlitz, and W. Wenzel -- Molecular docking / Garrett M. Morris and Marguerita Lim-Wilby. Subjects: Proteins--Structure--Computer simulation. Series: Methods in molecular biology, 1064-3745; 1215, Methods in molecular biology (Clifton, N.J.); 443.

Molecular simulations and visualization: University of Nottingham, Nottingham UK, 7-9 May 2014. LCCN: 2014497394 Main title: Molecular simulations and visualization: University of Nottingham, Nottingham UK, 7-9 May 2014. Published/Produced: Cambridge: Royal Society Of Chemistry, [2014] Description: 535 pages: color illustrations; 25 cm. ISBN: 9781782621430 (hbk.) 1782621431 (hbk.) LC classification: QP517. M65 M654 2014, Related names: Royal Society of Chemistry (Great Britain). Faraday Division, issuing body. Contents: Molecular simulations and visualization: introduction and overview / Jonathan D. Hirst, David R. Glowacki and Marc Baaden -- 3D molecular models of whole HIV-1 virions generated with cellPACK / Graham T. Johnson, David S. Goodsell, Ludovic Autin, Stefano Forli, Michel F. Sanner and Arthur J. Olson -- Innovative interactive flexible docking method for multi-scale reconstruction elucidates dystrophin molecular assembly / A. -E. Molza (and 8 others) -- A GPU-accelerated immersive audio-visual framework for interaction with molecular dynamics using consumer depth sensors / David R. Glowacki (and 9 others) -- Studying chemical reactivity in a virtual environment / Moritz P. Haag and Markus Reiher -- ExaViz: A flexible framework to analyse, steer and interact with molecular dynamics simulations / Matthieu Dreher (and 8 others) -- Virtual and augmented reality immersive molecular simulations: general discussion -- Visual analysis for space-time aggregation of biomolecular simulations / Thomas Ertl (and 5 others) -- Visualising intrinsic disorder and conformational variation in protein ensembles / Julian Heinrich, Michael Krone, Seán I. O'Donoghue and Daniel Weiskopf -- RiboVision suite for visualization and analysis of ribosomes / Chad R. Bernier (and 16 others) -- N Ras slows the rate at which a model lipid bilayer phase separates / Elizabeth Jefferys, Mark S. P. Sansom and Philip W. Fowler -- Density based visualization for molecular simulation / Dmitri Rozmanov, Svetlana Baoukina and D. Peter Tieleman -- Advanced visualization and visual analytics: general discussion -- GPU-accelerated analysis and visualization of large structures solved by molecular dynamics flexible fitting / John E. Stone, Ryan McGreevy, Barry Isralewitz and Klaus Schulten -- Visualising and controlling the flow in biomolecular systems at and between multiple scales: from atoms to hydrodynamics at different locations in time and space / Evgen Pavlov (and 5 others) -- Allosteric pathway identification through network analysis: from molecular dynamics simulations to interactive 2D and 3D graphs / Ariane Allain (and 5 others) -- Haptic-driven, interactive drug design: implementing a GPU-based approach to evaluate the induced fit effect / Athanasios Anthopoulos, Gaia Pasqualetto, Ian Grimstead and Andrea Brancale -- Accelerating electrostatic pair methods on graphical processing units to study molecules in supercritical carbon dioxide / John A. Baker and Jonathan D. Hirst -- A real-time proximity querying algorithm for haptic-based molecular docking / Georgios Iakovou, Steven Hayward and Stephen Laycock -- Computing power revolution and new algorithms: GP-GPUs, clouds and more: general discussion -- Immune Attack players perform better on a test of cellular immunology and self confidence than their classmates who play a control video game / Melanie Stegman --

Udock, the interactive docking entertainment system / Guillaume Levieux (and 5 others) -- Exploring genomes with a game engine / Jeremiah J. Shepherd (and 5 others) -- Methodologies for the analysis of instantaneous lipid diffusion in md simulations of large membrane systems / Matthieu Chavent (and 6 others) -- Rapid decomposition and visualisation of protein-ligand binding free energies by residue and by water / Christopher J. Woods (and 5 others) -- Applications and serious games: from docking to protein folding: general discussion -- Impressions by a dinosaur: summary of Faraday discussion 169: molecular simulations and visualization / Frederick P. Brooks, Jr. Subjects: Molecular dynamics--Congresses. Notes: "A general discussion on molecular simulations and visualization was held in Nottingham, UK on the 7th, 8th, and 9th May 2014"--Page 5.Includes bibliographical references. Series: Faraday discussions, 1359-6640; volume 169. Faraday discussions of the Chemical Society; v. 169.

Nobel prizes and nature's surprises, LCCN: 2013036833, Personal name: Norrby, Erling, author. Main title: Nobel prizes and nature's surprises / Erling Norrby, the Royal Swedish Academy of Sciences, Sweden. Published/Produced: New Jersey: World Scientific, [2013] Description: xviii, 451 p.: illustrations; 26 cm. ISBN: 9789814520980 (hard cover: alk. paper) 9789814520997 (pbk.: alk. paper) LC classification: QH315 .N666 2013, Contents: A magician of virology from Australia -- A divided Nobel Prize and a new era in immunology -- More Nobel Prizes in immunology -- Immunity, infections and transplantations -- Transgressing borders in science and scenes of life -- Making sense of hearing: art and science -- Unraveling the complexity of protein folding -- "It's so beautiful, you see, so beautiful!" -- Coda. Subjects: Allergy and Immunology--history. Nobel Prize. Biological Science Disciplines--history. Biomedical Research--history. History, 20th Century. Notes: Includes bibliographical references (pages 231-441) and index.

Plant genes, genomes, and genetics, LCCN: 2014028955, Personal name: Grotewold, Erich, author. Main title: Plant genes, genomes, and genetics / Erich Grotewold, Joseph Chappell, Elizabeth Kellogg. Published/Produced: Chichester, West Sussex, UK; Hoboken, NJ: John Wiley & Sons Inc., 2015. Projected pub date: 1502 Description: pages cm ISBN: 9781119998884 (cloth) 9781119998877 (pbk.) LC classification: QK981.4 .G76 2015, Related names: Chappell, Joseph, author. Kellogg, Elizabeth Anne, author. Contents: The plant genetic material -- The genome landscape: plastic and dynamic -- Transposable elements --

Centromeres, telomeres and chromatin -- Organelle genomes -- RNA -- The plant RNA polymerases -- Making mRNAs: control of transcription by RNP-II -- Controlling nuclear transcription -- Control of transcription factor function -- Small RNAs -- Chromatin and gene expression -- RNA processing and transport -- Fate of RNA -- Translation of RNA -- Protein folding and transport -- Protein degradation. Subjects: Plant molecular genetics. Plant gene expression. Genomics. Plants--genetics. Genomics. Plant Physiological Phenomena.RNA, Plant--genetics. Notes: Includes bibliographical references and index. Additional formats: Online version: Grotewold, Erich, author. Plant genes, genomes, and genetics Chichester, West Sussex, UK; Hoboken, NJ: John Wiley & Sons Inc., 2015 9781118539361 (DLC) 2014029985.

Protein aggregation and fibrillogenesis in cerebral and systemic amyloid disease, LCCN: 2012953928, Main title: Protein aggregation and fibrillogenesis in cerebral and systemic amyloid disease / J. Robin Harris, editor. Published/Produced: Dordrecht; New York: Springer, 2012, Description: xvii, 648 pages: illustrations (some color); 24 cm. ISBN: 9789400754157 (alk. paper) 9400754159 (alk. paper) LC classification: QP552.A45 P76 2012, Related names: Harris, James R., editor. Contents: Introduction and technical survey: Protein aggregation and fibrillogenesis -- Fibril formation by short synthetic peptides -- In vitrooligomerization and fibrillogenesis of amyloid-beta peptides -- Tau fibrillogenesis -- Prion protein aggregation and fibrillogenesisin vitro -- Alpha-synuclein aggregation and modulating factors -- Pathological self-aggregation of [beta](2)-microglobulin: A challenge for protein biophysics -- Islet amyloid polypeptide: Aggregation and fibrillogenesis in vitro and its inhibition -- Mechanisms of transthyretin aggregation and toxicity -- Fibrillogenesis of huntingtin and other glutamine containing proteins -- Aggregation and fibrillogenesis of proteins not associated with disease: a few case studies -- Experimental inhibition of peptide fibrillogenesis by synthetic peptides, carbohydrates and drugs -- Experimental inhibition of fibrillogenesis and neurotoxicity by amyloid-beta (abeta) and other disease-related peptides/proteins by plant extracts and herbal compounds -- Alzheimer's disease -- Modeling the polyglutamine aggregation pathway in Huntington's disease: from basic studies to clinical applications -- Parkinson's disease -- Human prion diseases: From kuru to variant Creutzfeldt-jakob disease -- Animal prion diseases -- [beta](2)-Microglobulin amyloidosis -- Systemic AA amyloidosis -- Familial amyloidotic polyneuropathy and transthyretin -- The challenge of systemic immunoglobulin light-chain amyloidosis (AL). Subjects: Amyloid beta-protein. Aggregation (Chemistry) Amyloidosis. roteins.Amyloidosis--physiopathology. Amyloid. Cell Aggregation. Plague, Amyloid. Protein Folding. Aggregation (Chemistry) Amyloid beta-protein. Amyloidosis. Proteins. Subject keywords: Medicine. Notes: Includes bibliographical references and index. Series: Subcellular biochemistry, 0306-0225; 65, Sub-cellular biochemistry; v. 65. 0306-0225.

Protein aggregation in bacteria: Functional and structural properties of inclusion bodies in bacterial cells, LCCN: 2013049564, Main title: Protein aggregation in bacteria: functional and structural properties of inclusion bodies in bacterial cells / edited by Silvia Maria Doglia, Marina Lotti. Published/Produced: Hoboken, New Jersey: Wiley, [2014] Description: xiv, 268 pages, 16 pages of unnumbered plates: illustrations (some color); 25 cm. ISBN: 9781118448526 (cloth) LC classification: QP552.C34 P75 2014, Related names: Doglia, Silvia Maria, editor of compilation. Lotti, Marina, editor of compilation. Contents: Fundamentals of protein folding / Vladimir N. Uversky -- Recruiting unfolding chaperones to solubilize misfolded recombinant proteins / Rayees U.H. Mattoo and Pierre Goloubinoff -- Osmolytes as chemical chaperones to use in protein biotechnology / Ario De Marco -- Inclusion bodies in the study of amyloid aggregation / Anna Villar-Piqu and Salvador Ventura -- Protein aggregation in unicellular eukaryotes / Marina Caldara, Joris Winderickx and Vanessa Franssens -- Structural properties of inclusion bodies / Antonino Natalello, Diletta Ami, Silvia Maria Doglia -- Residue-specific structure of proteins in inclusion bodies / Christian Wasmer, Marielle Wlti, Yongli Chen, Lei Wang -- Biomedical applications of bacterial inclusion bodies / Imma Ratera, Spela Peternel, Joaquin Seras-Franzoso, Olivia Cano-Garrido, Elena Garcia-Fruitcents, Rafael Cubarso, Esther Varquez, Jose Luis Corchero, Escarlata Rodriguez-Carmona, Jaume Veciana, Antonio Villaverde -- Aggregation of recombinant proteins: understanding basic issues to overcome production bottlenecks / Marina Lotti and Loredano Pollegioni -- Fusion to a pull-down module: designing enzymes to form biocatalytically active insoluble aggregates / Bernd Nidetzky. Subjects: Inclusion Bodies--physiology. Bacterial Proteins--chemistry. Bacterial Proteins--physiology. Bacterial Proteins--therapeutic use. Inclusion Bodies--chemistry. Industrial Microbiology--methods. Notes: Includes bibliographical references and index. Additional formats: Online version: Protein aggregation in bacteria Hoboken, New Jersey: Wiley, [2014] 9781118801499 (DLC) 2013051094, Series: Wiley series in protein and peptide science, Wiley series in protein and peptide science.

Protein and peptide folding, misfolding, and non-folding, LCCN: 2011044305, Main title: Protein and peptide folding, misfolding, and non-folding / edited by Reinhard Schweitzer-Stenner. Published/Created: Hoboken, N.J.: John Wiley & Sons, c2012. Description: xxii, 548 p., [24] p. of plates: ill. (some col.); 25 cm. Links: Cover image http://catalogimages.wiley.com/images/db/jimages/9780470591697.jpg, Table of contents only http://catdir.loc.gov/catdir/enhancements/fy1202/2011044305-t.html, Publisher description http://catdir.loc.gov/catdir/enhancements/fy1202/2011044305-d.html. Contributor biographical information http://catdir.loc.gov/catdir/enhancements/fy1205/2011044305-b.html; http://site.ebrary.com/id/10533977 ebrary ISBN: 97804705916970470591692, LC classification: QP551 .P75 2012, Related names: Schweitzer-Stenner, Reinhard. Summary: "This book provides an overview on what researchers have learned about unfolded peptides and how this knowledge facilitates the understanding of (a) the folding process, (b) the binding of ligands to receptor molecules, and (c) peptide self-aggregation. In this context, different experimental, theoretical, and computational concepts and approaches are introduced. This book can become a very useful addition for graduate-level courses on protein folding for the education of undergraduate and graduate students in research groups, which are exploring peptide self-aggregation for biomedical and biotechnological purposes"-- Provided by publisher. Subjects: Protein folding. Peptides. Notes: Includes bibliographical references and index. Series: Wiley series in protein and peptide science. Wiley series in protein and peptide science.

Protein conformational dynamics, LCCN: 2014930165 Main title: Protein conformational dynamics / Ke-li Han, Xin Zhang, Ming-jun Yang, editors. Published/Produced: Cham; New York: Springer, [2014] ©2014. Description: xii, 488 pages: illustrations (some color); 24 cm. Links: Publisher description http://www.loc.gov/catdir/enhancements/fy1405/2014930165-d.html, Table of contents only http://www.loc.gov/catdir/enhancements/fy1405/2014930165-t.html, ISBN: 9783319029696 (alk. paper)331902969X (alk. paper) LC classification: QP551 .P69576 2014, Related names: Han, Ke-Li, editor of compilation. Zhang, Xin, 1978- editor of compilation. Yang, Ming-jun, editor of compilation. Summary: "This book discusses how biological molecules exert their function and regulate biological processes, with a clear focus on how conformational dynamics of proteins are critical in this respect. In the last decade, the advancements in computational biology, nuclear magnetic resonance including paramagnetic relaxation enhancement, and fluorescence-based ensemble/single-molecule techniques have shown that biological molecules (proteins, DNAs and RNAs) fluctuate under equilibrium conditions. The conformational and energetic spaces that these fluctuations explore likely contain active conformations that are critical for their function. More interestingly, these fluctuations can respond actively to external cues, which introduces layers of tight regulation on the biological processes that they dictate. A growing number of studies have suggested that conformational dynamics of proteins govern their role in regulating biological functions, examples of this regulation can be found in signal transduction, molecular recognition, apoptosis, protein / ion / other molecules translocation and gene expression"-- Publisher's description. Contents: Protein folding simulations by generalized-ensemble algorithms -- Application of Markov state models to simulate long timescale dynamics of biological macromolecules -- Understanding protein dynamics using conformational ensembles -- Generative models of conformational dynamics -- Generalized spring tensor models for protein fluctuation dynamics and conformational changes -- The joys and perils of flexible fitting -- Coarse-grained models of the proteins backbone conformational dynamics -- Simulating protein folding in different environmental conditions -- Simulating the peptide folding kinetic related spectra based on the Markov state model -- The dilemma of conformational dynamics in enzyme catalysis: perspectives from theory and experiment -- Exploiting intrinsic flexibility in drug design -- NMR and computational methods in the structural and dynamic characterization of ligand-receptor interactions -- Molecular dynamics simulation of membrane proteins -- Free-energy landscape of intrinsically disordered proteins investigated by all-atom multicanonical molecular dynamics -- Coordination and control inside simple biomolecular machines -- Multi-state targeting machinery govern the fidelity and efficiency of protein localization -- Molecular dynamics simulations of F_1-ATPase -- Chemosensorial G-proteins-coupled receptors: a perspective from computational methods. Subjects: Proteins--Conformation. Biology. Protein Conformation. Notes: Includes bibliographical references and index. Additional formats: Online version: Protein conformational dynamics. Cham; New York: Springer, 2014 9783319029702 (OCoLC) 873822128 Series: Advances in experimental medicine & biology, 0065-2598; volume 805Advances in experimental medicine and biology; 805. 0065-2598.

Protein dimerization and oligomerization in biology, LCCN: 2012005555, Main title: Protein dimerization and oligomerization in biology / edited by Jacqueline M. Matthews. Published/Created: New York: Springer Science+Business Media; Austin, Tex.: Landes Bioscience, c2012. Description: xiv, 170 p.: ill.; 27 cm. ISBN: 9781461432289 (hardback) 1461432286 (hardback) LC classification: QP551.5. P74 2012, Related names: Matthews, Jacqueline M. Summary: "This volume has a strong focus on homo-oligomerization, which is surprisingly common. However, protein function is so often linked to both homo- and hetero-oligomerization and many heterologous interactions likely evolved from homologous interaction, so this volume also covers many aspects of hetero-oligomerization"--Provided by publisher. Contents: Machine generated contents note: Dimers, Oligomers, Everywhere.- The Detection and Quantitation of Protein Oligomerization -- Computational and Structural Characterisation of Protein Associations.- Death by Caspase Dimerization -- The Relationship between Oligomeric State and Protein Function -- Oligonucleotide Binding Proteins: The Occurrence of Dimer and Multimer Formation -- Homo and Heterodimerization in Transcriptional Regulation -- Oligomerization at the Membrane: Potassium Channel Structure and Function -- Implications of 3D Domain Swapping for Protein Folding, Misfolding and Function -- From artificial antibodies to nanosprings: The biophysical properties of repeat proteins. Subjects: Oligomerization. Protein Multimerization. Protein Folding. Notes: Includes bibliographical references and index. Series: Advances in experimental medicine and biology; v. 747Advances in experimental medicine and biology; v. 747. 0065-2598.

Protein engineering: design, selection and applications, LCCN: 2010004377, Main title: Protein engineering: design, selection and applications / Mallorie N. Sheehan, editor. Published/Produced: New York: Nova Science Publishers, c2013. ©2012 Description: xii, 221 p.: ill.; 24 cm. ISBN: 16261887859781626188785, LC classification: TP248.65.P76 P7436 2013, Related names: Sheehan, Mallorie N. Contents: Synthesizing mammalian cell adhesive surfaces using self-assembling fibronectin domains / Patricia Pereira and Christopher F. van der Walle -- Engineering thermobifida fusca cellulases: catalytic mechanisms and improved activity / Thu V. Vuong and David B. Wilson -- An empirical study for finding a robust ensemble of classifiers for protein classification / Loris Nanni and

Alessandra Lumini -- Revolutionary protein engineering using molecular display / Mitsuyoshi Ueda -- Chemoselective modification using aldehyde/ketone tags / Koichiro Kodama and Seketsu Fukuzawa -- Computer modeling of the promising HIV-1 entry inhibitors for structure-based anti-AIDS drug design: cyclophilin B and GP120 V3 loop / Alexander M. Andrianov and Ivan V. Anishchenko -- Molecular mechanisms of protein folding by mammalian ER chaperones and stress response / J. Geraldine Sandana Mala, Chellan Rose and Satoru Takeuchi -- Protein engineering of calcium binding proteins / Eugene A. Permyakov and Vladimir N. Uversky. Subjects: Protein engineering. Protein engineering--Methods. Notes: Includes bibliographical references (pages 198-204) and index. Series: Protein biochemistry, sythnesis, structure and cellular.

Protein folding and misfolding: shining light by infrared spectroscopy, LCCN: 2011937438, Main title: Protein folding and misfolding: shining light by infrared spectroscopy / Heinz Fabian, Dieter Naumann, editors. Published/Created: Heidelberg, Germany; New York: Springer, c2012. Description: xvi, 244 p.: ill. (some col.); 24 cm. ISBN: 9783642222290 (alk. paper) 3642222293 (alk. paper) 9783642222306 (e-ISBN) 3642222307 (e-ISBN) LC classification: QP551. P69582125 2012, Related names: Fabian, Heinz. Naumann, Dieter (Professor) Contents: Linked landscapes and conformational conversions: how proteins fold and misfold / Gareth J. Morgan and Sheena E. Radford -- A quantitative reconstruction of the amide I contour in the IR spectra of peptides and proteins: from structure to spectrum / Joseph W. Brauner and Richard Mendelsohn -- Millisecond-to-minute protein folding/misfolding events monitored by FTIR spectroscopy / Heinz Fabian and Dieter Naumann -- Watching dynamical events in protein folding in the time domain from submilliseconds to seconds: continuous-flow rapid-mixing infrared spectroscopy / Satoshi Takahashi and Tetsunari Kimura -- High-pressure vibrational spectroscopy studies of the folding, misfolding and amyloidogenesis of proteins / Roland Winter, Matthias Pühse, and Jonas Markgraf -- Dynamics of [alpha]-helix and [beta]-sheet formation studied by laser-induced temperature-jump IR spectroscopy / Karin Hauser -- Light-triggered peptide dynamics / Wolfgang Zinth and Josef Wachtveitl -- Time-resolved FTIR spectroscopy of pH-induced aggregation of peptides / John E.T. Corrie, Alex Perálvarez-Marín, and Andreas Barth -- Examining amyloid structure and kinetics with 1D and 2D infrared spectroscopy and isotope labeling / Lauren E. Buchanan, Emily B. Dunkelberger, and

Martin T. Zanni. Subjects: Protein folding. Proteins--Analysis. Infrared spectroscopy. Biochemistry. Protein Folding. Spectrophotometry, Infrared. Biochemistry. Infrared spectroscopy. Protein folding. Proteins--Analysis. Notes: "With 108 figures."Includes bibliographical references and index. Series: Biological and medical physics, biomedical engineering, 1618-7210.

Protein folding in silico: Protein folding versus protein structure prediction, LCCN: 2012942498, Main title: Protein folding in silico: protein folding versus protein structure prediction / edited by Irena Roterman-Konieczna. Published/Created: Oxford; Philadelphia: Woodhead Publishing Limited, 2012. Description: xxxi, 208 p.: col. ill.; 24 cm. ISBN: 9781907568176 (hbk.) 1907568174 (hbk.) LC classification: QP551 .P69582127 2012, Related names: Roterman-Konieczna, Irena. Summary: Protein folding in silico is broad in scope - starting with ab initio models and attempting to construct a mechanistic model on the basis of their predictions. The authors' computational model is expected to go beyond experimental studies, casting a new light on the complexities of biological phenomena -- Source other than Library of Congress. Subjects: Protein folding. Protein folding--Simulation methods. Protein Folding. Computer Simulation. Models, Theoretical. Notes: Includes bibliographical references and index. Series: Woodhead publishing series in biomedicine; no. 22Woodhead Publishing series in biomedicine; no. 22.

Protein homeostasis, LCCN: 2011023852, Main title: Protein homeostasis / edited by Richard I. Morimoto, Dennis J. Selkoe, Jeffery W. Kelly. Edition: 1st ed. Published/Created: Cold Spring Harbor, N.Y.: Cold Spring Harbor Laboratory Press, c2012. Description: viii, 349 p.: ill. (some col.); 26 cm. ISBN: 9781936113064 (hardcover: alk. paper) 1936113066 (hardcover: alk. paper) LC classification: QP551. P69594 2012, Related names: Morimoto, Richard I., 1952-Selkoe, Dennis J. Kelly, Jeffery W. Contents: I. Chemistry and biology of proteostasis -- Physical biochemistry of protein folding and misfolding / Michele Vendruscolo, Tuomas P. J. Knowles, and Christopher M. Dobson -- Integrating protein homeostasis strategies in prokaryotes / Axel Mogk, Damon Huber, and Bernd Bukau -- Cellular strategies of protein quality control / Bryan Chen, Marco Retzlaff, Thomas Roos, and Judith Frydman -- Integration of clearance mechanisms: the proteasome and autophagy / Esther Wong and Ana Maria Cuervo -- Aging as an event of proteostasis decline / Rebecca C. Taylor and Andrew Dillin -- The stress of protein misfolding: from single cells to multicellular organisms / Tali Gidalevitz, Veena Prahlad, and Richard I. Morimoto -- II. Protein quality control in cellular compartments -- Protein folding in the cytoplasm and the heat shock response / Martin Vabulas, Swasti Raychaudhuri, Manajit Hayer-Hartl, and F. Ulrich Hartl -- Protein folding and quality control in the ER / Kazutaka Araki and Kazuhiro Nagata -- Stress and its functional link to mitochondria: role in cell survival and death / Jyoti D. Malhotra and Randy Kaufman -- Quality control of mitochondrial proteostasis / Michael J. Baker, Takashi Tatsuta, and Thomas Langer -- III. Disorders of protein homeostasis -- Hepatic fibrosis and carcinogenesis in alpha-1-antitrypsin deficiency: a prototype for chronic tissue damage in gain-of-function disorders / David H. Perlmutter and Gary A. Silverman -- Alzheimer disease / Dennis J. Selkoe -- Huntington's disease / Steven Finkbeiner -- Proteostasis and movement disorders: Parkinson's disease and amyotrophic lateral sclerosis / Daryl A. Bosco, Matthew J. LaVoie, Gregory A. Petsko, and Dagmar Ringe -- Prions / David W. Colby and Stanley B. Prusiner -- Emergent properties of proteostasis in managing cystic fibrosis / William E. Balch, Daniela Roth, and Darren M. Hutt -- Protein misfolding and retinal degeneration / Radouil Tzekov, Linda Stein, and Shalesh Kaushal -- Chemical and Biological approaches for adapting proteostasis to ameliorate protein misfolding and aggregation diseases - progress and prognosis / Susan L. Lindquist and Jeffery W. Kelly. Subjects: Proteins--Metabolism. Homeostasis. Biological transport. Notes: "A subject collection from Cold Spring Harbor Perspectives in biology."Includes bibliographical references and index. Series: Cold Spring Harbor perspectives in biology, Cold Spring Harbor perspectives in biology.

Spin glasses and complexity, LCCN: 2012017289, Personal name: Stein, Daniel L. Main title: Spin glasses and complexity / Daniel L. Stein, Charles M. Newman. Published/Created: Princeton: Princeton University Press, c2013. Projected pub date: 1212 Description: p. cm. ISBN: 9780691147338 (pbk.) 0691147337 (pbk.) LC classification: QC176.8.S68 S74 2013, Related names: Newman, Charles M. (Charles Michael), 1946- Summary: "Spin glasses are disordered magnetic systems that have led to the development of mathematical tools with an array of real-world applications, from airline scheduling to neural networks. Spin Glasses and Complexity offers the most concise, engaging, and accessible introduction to the subject, fully explaining what spin glasses are, why they are important, and how they are opening up new ways of thinking about complexity. This one-of-a-kind guide to spin glasses begins by explaining the fundamentals of order and symmetry in condensed matter physics and how spin glasses fit into--and modify--this framework. It then explores how spin-glass concepts and ideas have found applications in areas as diverse as computational complexity, biological and artificial neural networks, protein folding, immune response maturation, combinatorial optimization, and social network modeling. Providing an essential overview of the history, science, and growing significance of this exciting field, Spin Glasses and Complexity also features a forward-looking discussion of what spin glasses may teach us in the future about complex systems. This is a must-have book for students and practitioners in the natural and social sciences, with new material even for the experts"-- Provided by publisher. Subjects: Spin glasses. Computational complexity. Notes: Includes bibliographical references and index. Series: Primers in complex systems.

The chaperonopathies: diseases with defective molecular chaperones: an introduction and guide to diseases in which chaperones play an etiologic-pathogenic role LCCN: 2013931840 Personal name: Macario, Alberto J. L., author. Main title: The chaperonopathies: diseases with defective molecular chaperones: an introduction and guide to diseases in which chaperones play an etiologic-pathogenic role / Alberto J. L. Macario, Everly Conway de Macario, Francesco Cappello. Published/Produced: Dordrecht; New York: Springer, [2013] ©2013 Description: xvi, 116 pages: illustrations (some color); 24 cm. ISBN: 9789400746664 (alk. paper) 9400746660 (alk. paper) LC classification: QP552.M64 M33 2013, Related names: Conway de Macario, Everly, author. Cappello, Francesco, author. Summary: This Brief provides a concise review of chaperonopathies, i.e., diseases in which molecular chaperones play an etiologic-pathogenic role. Introductory chapters deal with the chaperoning system and chaperoning teams and networks, HSP-chaperone subpopulations, the locations and functions of chaperones, and chaperone genes in humans. Other chapters present the chaperonopathies in general, including their molecular features and mechanistic classification into by defect, excess, or mistake. Subsequent chapters discuss the chaperonopathies in more detail, focusing on their distinctive characteristic. Subjects: Molecular chaperones. Diseases--Causes and theories of causation. Medical microbiology. Molecular Chaperones--analysis. Disease--etiology. Protein Folding. Diseases--Causes and theories of causation. Medical microbiology. Molecular chaperones. Notes: Includes bibliographical references. Series: Springer Briefs in biochemistry and molecular biology, 2211-9353SpringerBriefs in biochemistry and molecular biology. 2211-9353.

The molecules of life: physical and chemical principles, LCCN: 2012008865, Personal name: Kuriyan, John. Main title: The molecules of life: physical and chemical principles / John Kuriyan, Boyana Konforti, David Wemmer. Published/Created: New York: Garland Science, c2012. Projected pub date: 1208, Description: p.; cm. ISBN: 9780815341888 (alk. paper) Related names: Konforti, Boyana. Wemmer, David. Summary: "Key to making future advances in the areas of biochemistry and molecular medicine is a new generation of molecular biologists and biochemists who are able to harness the tools and insights of physics and chemistry to exploit the emergence of genomic and systems-level information in biology. The basic ideas of energy, entropy, equilibrium thermodynamics, transport processes and reaction kinetics are closely related to exciting issues in contemporary biology, such as protein folding, chaperones and prion diseases, DNA polymerase and ribosome fidelity, DNA recognition, drug design, signal transduction, ion channel function, motor protein action and the versatility of enzyme mechanism. Providing this physical chemistry and biochemical foundation is The "Molecules of Life", a new undergraduate textbook for undergraduate students majoring in biology or pre-med. It deepens our understanding of how life functions by illuminating the physical principles underpinning biological phenomena"--Provided by publisher. Subjects: Molecular Biology--methods. Biochemical Processes--physiology. Genomics--methods. Notes: Includes bibliographical references and index.

The protein folding problem and its solutions, LCCN: 2012046816, Personal name: Ben-Naim, Arieh, 1934- Main title: The protein folding problem and its solutions / Arieh Ben-Naim. Published/Created: Hackensack, New Jersey: World Scientific, c2013. Description: xix, 297 p.: ill. (some col.); 24 cm ISBN: 9789814436359 (hardcover: alk. paper) 9789814436366 (pbk.: alk. paper) LC classification: QP551. B34 2013, Subjects: Protein Folding. Protein Structure, Tertiary--physiology. Notes: Includes bibliographical references and index.

The structure of biological membranes, LCCN: 2011027921, Main title: The structure of biological membranes / editor, Philip L. Yeagle. Edition: 3rd ed. Published/Created: Boca Raton: CRC Press, 2012. Description: xii, 386 p.: ill.; 27 cm. ISBN: 9781439809570 (hardback: alk. paper) LC classification: QH601. S777 2012, Related names: Yeagle, Philip. Summary: "Biological membranes provide the fundamental structure of cells and viruses. Because much of what happens in a cell or in a virus occurs on, in, or across biological membranes, the study of membranes has rapidly permeated the fields of biology, pharmaceutical chemistry, and materials science. The Structure of Biological Membranes, Third Edition provides readers with an understanding of membrane structure and function that is rooted in the history of the field and brought to the forefront of current knowledge. The first part of the book focuses on the fundamentals of lipid bilayers and membrane proteins. Three introductory chapters supply those new to the field with the tools and conceptual framework with which to approach the state-of-the-art chapters that follow. The second part of the book presents in-depth analyses of focused subjects within the study of membranes, covering topics that include: Phase behavior of lipid bilayers as an isolated structure, Cholesterol's role in cell biology, Lateral organization of membranes, The role of membrane lipids in initial membrane protein folding, Membrane protein synthesis and assembly of oligomeric membrane proteins, Membrane protein stability with relationships to function and protein turnover, Membrane protein function using a transport protein, Interactions between membrane proteins and membrane lipids. A final chapter pulls together many of the topics, examining in detail the complexity inherent in the synthesis and assembly of lipids and proteins in mitochondrial membranes. With contributions from leading researchers, this completely revised and updated third edition reflects recent advances in the field of biological membranes."--Provided by publisher. Subjects: Membranes (Biology) Membrane proteins. Membrane lipids. Cell Membrane. Membrane Lipids. Membrane Proteins. Notes: Includes bibliographical references and index.

INDEX

#

A

B

C

D

E

F

G

H

I

J

K

L

M

N

O

P

T

U

V

W

Z